Lecture Notes
in Business Information Processing　　368

Series Editors

Wil van der Aalst
RWTH Aachen University, Aachen, Germany
John Mylopoulos
University of Trento, Trento, Italy
Michael Rosemann
Queensland University of Technology, Brisbane, QLD, Australia
Michael J. Shaw
University of Illinois, Urbana-Champaign, IL, USA
Clemens Szyperski
Microsoft Research, Redmond, WA, USA

More information about this series at http://www.springer.com/series/7911

Giovanni Meroni

Artifact-Driven Business Process Monitoring

A Novel Approach to Transparently Monitor
Business Processes, Supported by Methods,
Tools, and Real-World Applications

 Springer

Author
Giovanni Meroni ⓘ
Politecnico di Milano
Milan, Italy

ISSN 1865-1348   ISSN 1865-1356   (electronic)
Lecture Notes in Business Information Processing
ISBN 978-3-030-32411-7   ISBN 978-3-030-32412-4   (eBook)
https://doi.org/10.1007/978-3-030-32412-4

This book is a revised version of the PhD dissertation written by the author at: Politecnico di Milano, Italy. The original PhD dissertation is accessible at http://hdl.handle.net/10589/141243.

This Springer imprint is published by the registered company Springer Nature Switzerland AG
The registered company address is: Gewerbestrasse 11, 6330 Cham, Switzerland

In memory of my grandmother, Lidia

Preface

This book is a revised version of the PhD dissertation written by the author at the PhD School of Information Engineering of Politecnico di Milano (Italy). In 2019, the PhD dissertation won the CAiSE PhD Award, granted to outstanding PhD theses in the field of information systems engineering.

Traditionally, to monitor the execution of a business process, organizations rely on monitoring modules provided by Business Process Management Systems (BPMSs), which automate and keep track of the execution of processes [38]. While the adoption of a BPMS to monitor a single-party, fully-automated business process is straightforward, the same cannot be said for multi-party processes relying heavily on manual activities.

In fact, a BPMS requires explicit notifications to determine when activities that are not under its direct control are executed. This requires organizations to federate their BPMSs, a complex task that has to be performed whenever a new organization participates in the process. Also, when activities are not automated, human operators are responsible for manually sending notifications to the BPMS, a task that disrupts the operators' work and, as such, is prone to being forgotten or postponed.

To continuously and autonomously monitor multi-party processes involving non-automated activities, this book proposes a novel technique, named 'artifact-driven process monitoring.' This technique exploits the Internet of Things (IoT) paradigm to make the physical objects participating in a process 'smart.' Being equipped with sensors, a computing device, and a communication interface, such smart objects can then become self-aware of their own conditions and of the process they participate in, and exchange this information with the other smart objects and the involved organizations. In this way, it is possible for the monitoring infrastructure to stay in close contact with the process, and to cross the boundaries of the organizations.

To be aware of the process to monitor, instead of using activity-centric process models, usually adopted by BPMSs, smart objects rely on an extension of the Guard-Stage-Milestone (GSM) artifact-centric modeling language, named Extended-GSM (E-GSM). Normally, a BPMS expects the execution to rigidly adhere to the process model defined in advance. Therefore, whenever a deviation between the execution and the model is detected, a BPMS requires human intervention to resume process monitoring. E-GSM, on the other hand, treats the execution flow (i.e., dependencies among activities) in a descriptive rather than prescriptive way. Consequently, smart objects can detect violations during execution without interrupting the monitoring. Additionally, E-GSM can monitor if the physical objects evolve as expected while the process is executed. Finally, E-GSM provides constructs to determine, based on the conditions of the physical objects, when activities are started or ended.

This book also presents an approach to determine to what extent smart objects are suited to monitor a particular process, given their sensing capabilities. To relieve process designers from learning the E-GSM notation, and to allow organizations to reuse preexisting process models, a method to instruct smart objects given Business Process Model and Notation (BPMN) collaboration diagrams is also presented. Finally, a prototype of an artifact-driven monitoring platform, named SMARTifact, is developed and tested against both historical and live sensor data.

Acknowledgements

I would like to express my gratitude to all the people who helped me on the path towards achieving the PhD title. First and foremost, I would like to thank my supervisor Pierluigi Plebani, not only for giving me the opportunity, together with Prof. Luciano Baresi, to start the PhD school, but also for his support, comments, and useful feedback throughout my studies. I would also like to thank my tutor Prof. Barbara Pernici for her support.

I am also grateful to all my colleagues at Politecnico di Milano for providing a friendly and stimulating research environment, in particular Cinzia Cappiello, Monica Vitali, Florian Daniel, Xuesong Peng, Paolo Ravanelli, and Mattia Salnitri. Likewise, I would like to thank the research group in Information Business at WU Vienna, who hosted me as a visiting student, and in particular Prof. Jan Mendling and Claudio Di Ciccio. I am also very thankful to Prof. Marco Montali from the Free University of Bozen-Bolzano for the research collaboration carried out during my PhD work.

Additionally, I would like to thank the Italian project ITS 2020 and the Ministero dell'Istruzione dell'Università e della Ricerca (MIUR) for funding my PhD. I am also very thankful to the external reviewers, Prof. Massimo Mecella from Università di Roma La Sapienza and Prof. Barbara Weber from Technical University of Denmark, for their scrupulous review work and their useful comments. I would also like to express my gratitude to Prof. Selmin Nurcan and the CAiSE Committee for awarding me the prestigious CAiSE PhD Award.

Finally, I would like to thank my family for their love, understanding, support, and encouragement in undertaking the PhD school.

Contents

Chapter 1
Introduction

1.1 Motivations

To remain competitive, today's organizations are required to promptly satisfy the ever-changing needs of their customers. To do so, organizations must become more flexible and open to changes and opportunities. Such a flexibility is supported by the *servitization* paradigm [127]. Instead of owning corporate assets, be them physical goods, human resources, or business activities, an organization establishes contracts with other organizations, named service providers, that, in exchange for a periodic fee, grant the use of such assets together with value-added services (e.g., maintenance).

Servitization allows organizations to easily and inexpensively acquire new assets whenever new requirements arise, since no upfront investment is required. Similarly, assets can be alienated as soon as they become no longer needed by simply resolving the contract. Additionally, assets can be replaced with newer ones before they become outdated with no costs other than the periodic fee, thus avoiding obsolescence. For example, instead of having its own logistics division, a chemical company may rely on a international logistics company to deliver its products to its customers worldwide. The logistics company, in turn, may externalize each leg composing a shipment to a transportation company operating on a specific means of transport (e.g., trains, trucks, etc.). Finally, the transportation companies may choose to lease the vehicles composing their fleet, rather than buying them, from the manufacturers of these vehicles.

As a consequence of servitization, many business processes that were internal now cross the boundaries of single organizations, thus becoming *multi-party*. This also affects the goods that participate in the process, that are now manipulated and possibly altered by multiple organizations when the process is executed. Additionally, the identity of the service providers involved in these multi-party processes is subject to frequent changes. In fact, an organization may rely on multiple service providers, each one allocated on different process instances. Alternatively, an organization may define short-term contracts with service providers, in order to assign its processes to the one that best fits its current needs.

© Springer Nature Switzerland AG 2019
G. Meroni: Artifact-Driven Business Process Monitoring, LNBIP 368, pp. 1–12, 2019
https://doi.org/10.1007/978-3-030-32412-4_1

Despite the previously mentioned advantages it brings, servitization also requires organizations to trust each other. In particular, it cannot be taken for granted that processes are executed as planned. As organizations can enforce the execution of only those activities under their responsibility (i.e., internal) to adhere to the model, they have no control on the portions of the process that are externalized. As a consequence, unexpected deviations in the execution of the process and in the usage of the goods may occur. If not promptly identified and notified to the other organizations, such deviations may cause delays, coordination problems, and consequently dissatisfaction for the customers.

The main goal of this book is to reliably monitor the execution of (partially) outsourced business processes, as this is a critical aspect to coordinate organizations and ensure that contracts are fulfilled. To monitor a process in a reliable way, activities composing the process should be identified as running if and only if they are actually being performed. Additionally, discrepancies between the expected execution of the process and the actual one should be detected as soon as they occur. This would allow organizations to promptly take countermeasures, and possibly charge the one responsible for the deviation a penalty. Finally, the process should be monitorable also when the identity of the organizations is subject to frequent changes, possibly even while the process is being executed.

As shown in the top portion of Figure 1.1, to monitor the execution of a process, organizations traditionally rely on a Business Process Management System (BPMS). A BPMS is a software component aimed at automating the execution of a process [38] and, by doing so, it keeps track of each process being executed. However, BPMSs generally target single-party business processes, and as such they present several shortcomings when it comes to monitor multi-party processes. Also, they expect the execution to rigidly adhere to the process defined in advance, and are usually unable to automatically continue the execution of a process once a deviation occurs. Finally, a BPMS requires operators to send notifications whenever activities are initiated and completed, a task which can difficultly be enforced, especially in multi-party processes.

To monitor the usage of the goods, different techniques have been put in place. The most recent ones exploit the Internet of Things (IoT) paradigm, that turns physical goods into smart objects. This makes possible to uniquely identify physical objects across organizations. Thus, organization can know which physical goods participate in each process execution. If smart objects are equipped with sensors, it becomes also possible to know their physical conditions and how they are influenced by the interaction with the organizations. In this case, the smart object transmits sensor data to its owner, who can then make these data accessible to the other organizations. It is worth noting that, in all these cases, smart objects are not aware of the process in which they are participating. Therefore, organizations have to manually correlate information on the process with data coming from the smart objects.

To sum up, to maximize the advantages brought by the servitization paradigm, organizations need to monitor multi-party processes. In particular, organizations need to know whenever their goods are incorrectly manipulated, and if the outsourced portions of their processes are not executed as agreed. Violations should

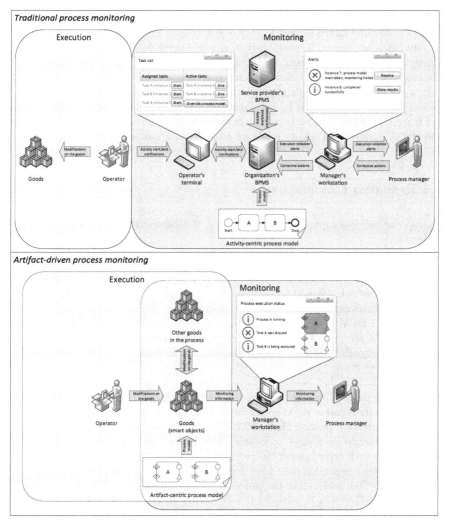

Fig. 1.1: Traditional approach to monitor a business process (top), compared to artifact-driven process monitoring, discussed in this book (bottom).

be detected while the process is run, and not upon completion. This way, the sooner violations are detected, the faster organizations can adapt the process and react to them. Also, the burden of a monitoring infrastructure, both in terms of deployment and configuration, and in interaction with the human operators, should be minimized. In particular, organizations should be free to make new contracts, or resolve old ones, without having to manually adapt and federate their monitoring infrastructures. Human operators should not interrupt their work to send notifications on what they are currently doing.

This book addresses the previously mentioned needs by proposing a novel technique to monitor the execution of multi-party processes, named artifact-driven process monitoring. Instead of monitoring a process on the organizations' premises with a BPMS, this technique exploits the IoT to perform the monitoring directly on the smart objects owned by the organizations, a shown in the bottom portion of Figure 1.1. In the next sections, the research issues that are tackled in this book, and the contributions that are made for each issue, are extensively discussed.

1.2 Research Challenges

To improve monitoring multi-party processes, the following main issues have to be addressed:

- **C1: Having visibility on the outsourced process portions.** To monitor their processes, organizations resort to deploying a BPMS. A BPMS is driven by a process model, a formal description of the process being executed, expressed in terms of business activities and their dependencies. The monitoring components of today's BPMSs are good at overseeing the execution of business processes, as long as the processes can be completely automated and confined within a single party. BPMSs also provide dashboards to inform the process owner of the current status, bottlenecks, and possible alerts.

  Unfortunately, when portions of the process are outsourced to other organizations, the process becomes multi-party and, a such, a BPMS becomes no longer able to fully monitor it. As a BPMS expects the process to be under control of the organization owning that BPMS, it expects information on the execution of activities to be always accessible. This is not the case in multi-party processes, as each organization by default has no access on the information owned by the other organizations. Therefore, such a BPMS cannot directly monitor how the other portions of the process, carried out by the other organizations, are executed. As such, a complete monitoring on the whole process cannot be directly achieved.

  This limitation is traditionally addressed by federating the BPMSs of the organizations participating in the process. This way, each BPMS can exchange information on its process portion to the other BPMSs, thus obtaining a global view on how the whole process is being executed. Another solution consists in deploying a centralized BPMS, that can be used by all the collaborating organizations. This solution requires either one of the collaborating organizations, or a third party, to put in place the centralized BPMS.

  While these solutions can be reasonable for long-term collaborations among organizations, they lack flexibility when it comes to frequently changing collaborations. In particular, whenever a new organization is involved in the process, or an organization no longer collaborates, the underlying infrastructure must be heavily reconfigured. In case of federated BPMSs, the messages that each

BPMS has to send and receive have to be redefined. In case of a centralized BPMS, its access rights and authorizations must be tailored to the collaborating organizations. Therefore, these solutions are unfit to support the servitization paradigm, which requires flexible, ever changing collaborations.

- **C2: Monitoring non-compliant executions autonomously and continuously.** As previously mentioned, a BPMS expects to have full control on the process being executed. As a consequence, it tries to enforce the process to be performed as defined in the process model. Therefore, when a deviation in the execution of a process occurs (e.g., an activity that is not supposed to be executed begins), a BPMS can no longer trust the process model. In such a case, the typical behavior of a BPMS would be to simply raise an exception and either stop the process, or ask a user to specify how the process should continue its execution.

This is highly advisable when the BPMS is used to drive the execution of a process. However, when it comes to monitor a process whose execution cannot be enforced, this behavior becomes undesired. This is the case of multi-party processes and, to a lesser extent, processes relying on activities that are not automated.

Being a multi-party process distributed among different stakeholders, the BPMS can enforce the execution of only those process portions carried out by the organization who owns the BPMS. As a consequence, the execution of the other portions, carried out by the other organizations, can only be passively observed.

In case of activities that are non automated, an operator may choose to ignore the instructions given by the BPMS, and arbitrarily decide when and if such activities should be executed. Since operators are taught to follow the directives of the BPMS, and possibly sanctioned if they disobey, this issue is not so frequent inside an organization. However, no guarantee can be taken that operators will always execute these activities according to the process model.

In both of these cases, the process could continue running even after a violation occurred, causing the monitoring to be unreliable. In particular, if the BPMS stops monitoring the execution of the process, no activity following the violation could be tracked. Even if the BPMS continues to monitor the process, it would be unable to detect any further violation until someone tells it how the process should continue its execution.

A conventional approach to solve this issue would be tuning the BPMS not to stop in case of violations, and resorting to conformance checking tools to detect which portion of the process is affected by each violation. These tools usually resort to process log mining techniques: they analyze the so-called process traces, information collected by the BPMS on when activities were started or finished, to derive a model of the process based on its execution. Such a model is then compared with the original one, and the (possible) differences in the two models indicate which violations occurred and which portions of the process were affected.

The main disadvantage of these conformance checking tools is that they usually work post-mortem, i.e., once the process completes its execution. As a consequence, stakeholders would not be able to detect violations as soon as they materialize. Therefore, they would not be able to estimate the effects of a violation on the outcome of the process, and immediately take countermeasures.

- **C3: Correlating the usage of the goods with the process they participate in.** Traditionally, business processes have been formalized with activity-centric languages. These languages mainly describe the process in terms of activities that are executed, and control-flow dependencies among these activities. However, a business process is also responsible for manipulating physical or virtual goods, named artifacts. Such artifacts may participate in the whole process, or only to specific portions of it. Consequently, it is also possible to describe a business process in terms of artifacts participating in that process, and changes in their characteristics that are expected to occur when the process is executed. Artifact-centric languages allow to formalize the process this way. Being able to monitor both activities and artifacts allows to have a more accurate idea on how the process is being executed. For example, it may happen that two process executions may lead to one artifact with different characteristics, even though the same activities were executed in the same order. In this case, monitoring a process just in terms of activities and their dependencies would not detect such a deviation. Alternatively, even though two process executions cause artifacts to evolve in the same way, such a result may be obtained by executing different activities. In this case, such a deviation would not be detected if the monitoring focuses only on the artifacts and their evolution.

When a BPMS is adopted to monitor a process, it requires the process to be modeled with an activity-centric language. This allows the process to be monitored in terms of activities being executed, resources being allocated, and time spent. However, it does not allow to monitor how artifacts change during execution. Therefore, to be aware of the characteristics of the artifacts, and how they change when the process is executed, organizations have to also put in place solutions to track and monitor these artifacts. To do so, these solutions exploit sensor networks or, more recently, the IoT paradigm. For example, in the logistics domain, monitoring solutions to track and trace the goods being shipped are widely adopted.

Even though solutions to monitor artifacts have been available for a long time, even before Business Process Management (BPM), they either lack the concept of business process, or the process they rely on is partially or fully hard coded into the solution. As such, changes in the artifacts have to be manually correlated with the process being executed. Also, when the process changes, the solution has to be heavily reconfigured. Therefore, they lack the flexibility to cope with ever-changing business processes.

To represent information on the artifacts and their characteristics, some activity-centric languages, like Business Process Model and Notation (BPMN) 2.0, introduce constructs to define which artifacts participate in a process, and which activities use, produce, or alter those artifacts. However, the evolution of the artifacts is only implicitly defined in the process model, based on the dependencies among activities. As a consequence, when the process is monitored, it is not verified if the artifacts evolve as expected.

- **C4: Detecting without human interactions when activities are executed.** To reliably monitor a process, the activation and termination of all the activities composing the process should be recorded. When a BPMS is introduced, it can autonomously detect when fully or partially automated activities are executed. As the BPMS is responsible for executing fully automated activities, it implicitly knows when they start and when they finish. Similarly, in order to execute partially automated activities, the user has to interact with the BPMS, typically by filling in a form. As such, the BPMS can easily identify when the interaction starts and how long it lasts. However, when activities are not automated, i.e., they are performed by humans who do not need to interact with the BPMS to execute them, being able to determine when they are executed becomes quite challenging.

  As no interaction with the BPMS occurs when activities are not automated, the only way to inform the BPMS on their execution would be to manually send a notification. Usually, a BPMS shows which activities an operator should execute in the form of a task list. Operators then, by flagging items in the task list, notify the BPMS when activities start and when their execution is complete. However, this causes operators to interrupt their work. As such, it could be easily forgotten or simply postponed, thus negatively affecting the reliability of the monitoring.

1.3 Research Questions

To tackle the problems and issues described in the previous section, the following main research questions are defined:

- **RQ1:** How can process portions that are executed by different organizations be monitored? This question concerns challenge C1, and can be furtherly expanded as follows:

  - **RQ1.1:** How can each organization be aware of the execution of those process portions being carried out by other organizations?
  - **RQ1.2:** Can a monitoring platform cross the boundaries of an organization?
  - **RQ1.3:** Can the IoT paradigm be used to decentralize process monitoring?

- **RQ2:** How can organizations autonomously and continuously monitor business processes? This question concerns C2, and can be expanded as follows:

  - **RQ2.1:** How can incorrect executions of a process be detected as soon as they happen?
  - **RQ2.2:** How can a process be monitored even after a violation occurred?
  - **RQ2.3:** Is it possible to determine the impact of a violation on the execution of the process?

- **RQ3:** How can the conditions of the goods participating in a process be monitored? This question concerns C3, and can be expanded as follows:

– **RQ3.1:** Can the IoT paradigm allow physical goods to become self-aware of their conditions?
– **RQ3.2:** Is it possible to determine if the goods are correctly manipulated? Can the expected evolution of the goods participating in a process be formalized? If so, is it possible to detect and promptly report violations in such an evolution?
– **RQ3.3:** How can the identity of the goods participating in a process execution be specified?
– **RQ3.4:** Which are the dependencies among the conditions of the goods and the execution of a process?

- **RQ4:** How can the execution of the activities be autonomously detected? This question concerns C4, and can be expanded as follows:

   – **RQ4.1:** Is it possible to detect when the activities start or end without bothering operators?
   – **RQ4.2:** Can the goods manipulated during the execution of a process be used to determine when the activities are executed?
   – **RQ4.3:** Can the IoT paradigm be used to detect when the activities start or end?
   – **RQ4.4:** Is it possible to assess to which extent the monitoring infrastructure is suited to monitor activities before the monitoring takes place? It is possible to automatically provide suggestions on how to improve the monitoring infrastructure?

1.4 Major Contributions

With respect to the research challenges and research questions, the contributions of this book are the following:

- **Artifact-driven process monitoring, a novel approach to monitor business processes.** Artifact-driven process monitoring relies on the changes in the state of the artifacts participating in a process to detect when activities are executed. This way, manual notifications are no longer required to determine when non-automated activities are started and completed. Additionally, artifact-driven process monitoring exploits the IoT paradigm to turn physical artifacts into smart objects, equipped with sensors, computing devices, and communication interfaces. This way, such smart objects can autonomously infer their own state from sensor data.
This contribution addresses RQ1.1, RQ1.3, RQ3.1, RQ3.4, RQ4.1, RQ4.2, and RQ4.3. Publications related to this contribution are *(i) Giovanni Meroni. Integrating the Internet of Things with Business Process Management: A Process-aware Framework for Smart Objects. In Proceedings of the CAiSE'2015 Doctoral Consortium* [82], *(ii) Luciano Baresi, Giovanni Meroni, and Pierluigi Plebani. A GSM-based Approach for Monitoring Cross-Organization Business Processes Using Smart Objects. In BPM 2015 Workshops Proceedings* [14], and

(iii) Giovanni Meroni, Luciano Baresi, Marco Montali, and Pierluigi Plebani. Multi-party business process compliance monitoring through IoT-enabled artifacts. In Information Systems, Volume 73 [83].

- **Extended-GSM (E-GSM), an extension of Guard-Stage-Milestone (GSM) to autonomously monitor business processes.** With respect to traditional activity-centric languages, such as BPMN, in E-GSM control-flow dependencies are descriptive rather than prescriptive. By doing so, when activities are executed, an E-GSM engine simply verifies if such control-flow dependencies are satisfied and, if not, it autonomously flags the activity as not executed at the right time. This way, the process continues to be monitored even after a violation occurs, and no human intervention is required. Additionally, E-GSM allows to define conditions on external data to determine when activities are started, completed, and if something went wrong when they were executed.

  This contribution addresses RQ2.1, RQ2.2 and RQ2.3, and it has been presented in *Luciano Baresi, Giovanni Meroni, and Pierluigi Plebani. On Handling Business Process Anomalies through Artifact-based Modeling. In Proceedings of the CAiSE'16 Forum* [15].

- **A method to derive E-GSM models from BPMN collaboration diagrams.** As mentioned in the previous point, E-GSM allows to autonomously monitor business processes. However, it requires process designer to learn new constructs, and a completely new modeling paradigm. To relieve designers from getting acquainted with E-GSM, and to allow organizations to reuse preexisting process models, a method to transform such process models into E-GSM is proposed. Starting from BPMN collaboration diagrams, this method guides the process designer in enriching such model with information on the artifacts, and in extracting the portions of the process that are relevant for each artifact. Then, each process portion is automatically translated into two E-GSM models, representing the activities composing that portion, and the expected changes in the characteristics of the artifact.

  This contribution addresses RQ3.2, and it has been presented in *(i) Luciano Baresi, Giovanni Meroni, and Pierluigi Plebani. Using the Guard-Stage-Milestone Notation for Monitoring BPMN-based Processes. In BPMDS EMMSAD 2016 Proceedings* [16], and in *(ii) Giovanni Meroni, Luciano Baresi, Marco Montali, and Pierluigi Plebani. Multi-party business process compliance monitoring through IoT-enabled artifacts. In Information Systems, Volume 73* [83].

- **Criteria to dynamically bind smart objects participating in the same process execution.** When information on the artifacts are introduced in the process model, the identity of these artifacts may no be known yet. Oftentimes, it is known only once the process is executed. In some cases, the identity of one or more artifacts may even change during a process execution. However, to exchange information, every smart object needs to known the identity of the other ones that participate in the same process execution. To provide this information once the monitoring started, criteria to dynamically bind and unbind smart objects to a specific process execution are defined. Also, mechanisms to automatically derive such criteria from a BPMN collaboration diagram are introduced.

This contribution addresses RQ3.3. Publications related to this contribution are *(i) Giovanni Meroni, Claudio Di Ciccio, Jan Mendling. Artifact-driven Process Monitoring: Dynamically Binding Real-world Objects to Running Processes. In Proceedings of the CAiSE'17 Forum and Doctoral Consortium* [86], and as *(ii) Giovanni Meroni, Claudio Di Ciccio, Jan Mendling. An Artifact-Driven Approach to Monitor Business Processes Through Real-World Objects. In Proceedings of ICSOC 2017* [85].

- **SMARTifact, an artifact-driven monitoring platform.** With respect to other monitoring platforms exploiting the IoT paradigm, that rely on smart objects simply to collect and forward sensor data, SMARTifact relies on a lightweight E-GSM engine that can directly run on top of smart objects, as long as they are equipped with a Single-board Computer (SBC) like the Intel Galileo[1] or the Raspberry Pi[2]. This way, smart objects become self-aware on the process they participate in. They can also exchange information on their state with the other smart objects participating in the same process execution, in a Machine-to-Machine (M2M) fashion.

  This contribution addresses RQ1.2, and it has been presented in *(i) Luciano Baresi, Claudio Di Ciccio, Jan Mendling, Giovanni Meroni, and Pierluigi Plebani. mArtifact: an Artifact-driven Process Monitoring Platform. In Proceedings of BPM 2017 Demo Track and Dissertation Award* [11] (This publication referred to SMARTifact with its former name, mArtifact), and in *(ii) Giovanni Meroni, Luciano Baresi, Marco Montali, and Pierluigi Plebani. Multiparty business process compliance monitoring through IoT-enabled artifacts. In Information Systems, Volume 73* [83].

- **A technique to formalize, assess and improve the capabilities of smart objects.** For the monitoring to be reliable, the characteristics of the artifacts defined in the process model must be observable. In particular, a smart object must be equipped with sensors capable of measuring the characteristics of the artifact embodied by the smart object. However, manually determining if the smart object are suited to monitor a process can be cumbersome. To do so, the process model must be inspected, and the characteristics of the referenced artifacts must be compared with the smart objects embodying the artifacts. To ease such an inspection, a technique to formalize the capabilities of smart objects, and to quantify the *monitorability* of a process, that is, to which extent a process can be monitored given a set of smart objects, is proposed. This technique can also provide suggestions to the process designer on how to modify the process model, and to equip the smart objects with new sensors, in order to improve the monitorability of the process.

  This contribution addresses RQ4.4, and it has been presented in *Giovanni Meroni and Pierluigi Plebani, Artifact-Driven Monitoring for Human-Centric Business Processes with Smart Devices: Assessment and Improvement. In Proceedings of BPM Forum 2017* [88].

---

[1] See https://software.intel.com/en-us/iot/hardware/galileo.

[2] See https://www.raspberrypi.org.

1.5 Book Structure

This book is organized as follows:

- Chapter 2 introduces the related work and background with respect to the main concepts involved in this book. Firstly, current state-of-the-art solutions to monitor the execution of business processes are introduced. Then, the main families of declarative languages are presented, as well as techniques to shift to declarative languages given imperative process models. Finally, an extensive introduction on the IoT paradigm, the enabling technologies, and the synergies with BPM is presented.
- Chapter 3 describes the artifact-driven approach to monitor multi-party processes. In addition, the reference architecture of an artifact-driven monitoring platform is introduced.
- Chapter 4 describes E-GSM, extension of the GSM artifact-centric language, used to drive the monitoring platform. This chapter shows how E-GSM can be used to autonomously monitor multi-party processes, by automatically dealing with violations between the model and the execution. The increase in expressiveness that E-GSM brings is also demonstrated by comparing it against imperative process models. A discussion on how E-GSM can be used to determine the impact of a violation on the outcome of the process concludes this chapter.
- Chapter 5 presents a method to automatically transform BPMN collaboration diagrams into E-GSM models suited to drive the monitoring platform. This chapter also shows how the E-GSM models resulting of this method can monitor both the activities being executed and the evolution of the artifacts. Additionally, a formal proof of correctness of the transformation is provided.
- Chapter 6 provides a technique to quantify the fitness of smart objects and process models to monitor a specific business process. Firstly, two ontologies are introduced to describe in a formal way the capabilities of the smart objects owned by each organization. Then, a quantitative metric, named *monitorability*, is introduced to assess to which extent a process can be monitored. Finally, given a process model, by querying the ontologies, the monitorability of such a process can be automatically quantified. Additionally, suggestions on how to modify the smart objects or the process model to improve the monitorability can be automatically obtained by querying the ontology.
- Chapter 7 describes the experiments that were made to evaluate the effectiveness of the artifact-driven process monitoring approach. SMARTifact, a monitoring platform implementing the reference architecture introduced in Chapter 4, is firstly validated against both processes and historical data from a company operating in the logistics domain. A smart object prototype running SMARTifact is then built and employed to monitor a live shipping processes.
- Chapter 8 draws the conclusions for this book, and outlines possible future directions.

Chapter 2
Related Work

To better understand the advantages that artifact-driven process monitoring brings, this chapter firstly explains what process monitoring is, and how it positions with respect to the BPM lifecycle. Then, methods and techniques related to process monitoring are presented. For each of them, current research work in that direction is surveyed, and their potentials, as well as their current limitations, are discussed.

Artifact-driven process monitoring relies on the E-GSM declarative process modeling language to represent the process to be monitored. To justify this choice, the declarative modeling paradigm is briefly introduced. The differences between this paradigm and the imperative one are also underlined, together with the advantages and disadvantages that declarative modeling brings. In addition, three of the most famous declarative languages, namely Declare, Guard-Stage-Milestone (GSM), and Case Management Modeling Notation (CMMN), are also presented. Also, research work in transforming imperative process models into declarative ones, in order to ease the adoption of this paradigm, is surveyed.

To autonomously detect when activities are executed and if they are incorrectly performed, artifact-driven process monitoring relies on the IoT. To this aim, a brief introduction on the IoT and its enabling technologies is presented. In addition, an overview of ongoing research work to formalize the characteristics of an IoT-based solution with ontologies is provided. Finally, the challenges that the integration of the IoT with BPM poses are discussed, and the research work going in this direction is surveyed.

2.1 Business Process Monitoring

According to Weske in [132], the lifecycle of BPM has a cyclic behavior, and is composed of four main phases, as shown in Figure 2.1:

- **Design and analysis**. Firstly, the business process is identified, and a model representing the process is manually designed and/or automatically elicited. The obtained process model is then validated by assessing its formal correctness, simulating all the possible executions, estimating costs, execution time

© Springer Nature Switzerland AG 2019
G. Meroni: Artifact-Driven Business Process Monitoring, LNBIP 368, pp. 13–36, 2019
https://doi.org/10.1007/978-3-030-32412-4_2

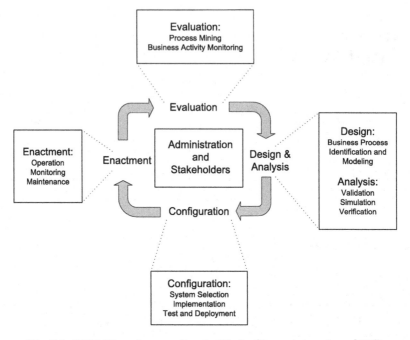

Fig. 2.1: BPM lifecycle according to Weske (figure taken from [132]).

and resources allocation. All these analysis are carried out based uniquely on the model, without executing the process. In addition, the model is manually inspected to verify that all the relevant aspects characterizing the process are captured in the model.

- **Configuration**. Once the process model is defined and validated, the process is implemented. This can be done in different ways. The process model can be refined to become executable by a BPMS, a software component responsible for coordinating the execution of the process according to the model. Alternatively, an ad-hoc software that behaves according to the process model can be developed. Finally, the process model can be used to derive a set of policies or guidelines that human operators should follow. It is worth noting that, depending on the way the process is implemented, the execution of the process can be completely, partially or not be carried out automatically and autonomously. Generally speaking, whenever human operators are involved to execute activities, the process is not completely automated.

- **Enactment**. Once the process is implemented, it can be executed. While the process is running, it is monitored to identify when actual executions take place (i.e., an instance of the process is created) and terminate, when activities composing the process are started and concluded, and if exceptions during the execution of the process or a specific activity occur. This information is usually collected and stored in log files. In case the process is (partially) automated,

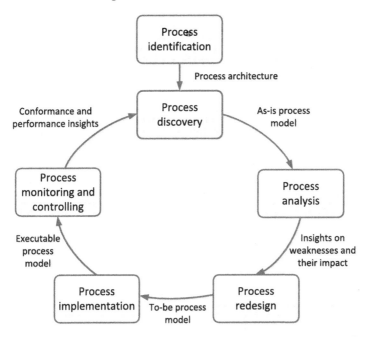

Fig. 2.2: BPM lifecycle according to Dumas et al. (figure taken from [38]).

the BPMS or ad-hoc software makes sure that activities are executed at the right time.

- **Evaluation**. Once the process is enacted, the real efficiency and effectiveness of the process is assessed. For instance, the average time spent running each activity can be determined, as well as the idle time. Possible bottlenecks in the execution and allocation of activities can be identified. Additionally, it can be determined to which extent the process model reflects the actual execution of the process. Such an information can then be used as an input for the Design and Analysis phase, thus completing the cycle.

Dumas et al. in [38] also describe the lifecycle of BPM with a cyclic behavior, as shown in Figure 2.2. However, they propose a slightly different classification of the phases characterizing such a lifecycle.

- The *design and analysis* phase is broken into four phases carried out in sequence:

  - **Process identification**. This phase aims at identifying the existence of one or more business processes relevant for a specific business problem, and their dependencies.
  - **Process discovery**. Once the existence of a process is determined, this phase determines how the process is performed and capturing such an information in a process model.

- **Process analysis**. After the process model is defined, this phase statically analyzes such a model to determine performance issues and opportunities for improvement.
- **Process redesign**. Based on the results of the previous phase, the process model is altered to avoid the issues and benefit from the opportunities that were identified.

- The *enactment* and *evaluation* phases are merged into a single phase named **process monitoring and controlling**. A possible reason for such a merge is the tight integration among the collection of process execution data and the assessment of the actual efficiency and effectiveness. Usually, process execution is assessed post-mortem (i.e., after the execution ends), and as such it is done after the process is enacted. However, techniques to asses the execution at runtime also exist, thus allowing to carry out such a phase in parallel to the enactment one.

As the main focus of this book is on business process monitoring, from this point on only the *process monitoring and controlling* phase will be discussed in detail.

In [1], current methods and techniques related to BPM are surveyed. Among the techniques supporting the *process monitoring and controlling* phase, the ones relevant to monitor a process are the following:

- **Event data logging** consists in collecting events related to the execution of a specific process instance. Such events can refer to the activities being executed, and to the artifacts (i.e., the physical or virtual objects manipulated by the process) and the resources (i.e., the human operators or software components responsible for executing activities) interacting with the process. Once collected, events are typically stored in so-called event logs. Given a running system, *event data logging* produces event data.
- **Business Activity Monitoring (BAM)**, which [1] simply calls "monitoring", consists in collecting real-time information on the activities being executed, such as response time and failure rate. Such an information is then used to provide Key Performance Indicators (KPIs) to the stakeholders, thus allowing them to estimate how well activities are performed. With respect to other techniques, *BAM* relies on information related to the single activities, without taking the process model into consideration. Given a running system, *BAM* produces KPIs.
- **Conformance checking** consists in verifying if the modeled behavior of the process differs from the observed one. To do so, event data is replayed and compared with the process model. Whenever an inconsistency is detected, the affected portion of the process model and the event causing such an inconsistency are determined. Given a process model and event data, *conformance checking* produces conformance-related diagnostic information.
- **Runtime performance analysis** consists in extracting performance information on the processes being executed. This way, bottlenecks or resource allocation problems can be identified. *Runtime performance analysis* differs from *BAM* since dependencies among activities are taken into account. It also differs from techniques performed during the design and analysis phase, such as

design-time analysis and simulation, as it uses real data coming from actual process executions, rather than estimations. Given a process model and event data, *runtime performance analysis* produces performance-related diagnostic information.

Even though not explicitly mentioned in [1], another technique related to BPM is **compliance checking**. As discussed by Reichert and Weber in [112], *compliance checking* consists in verifying that semantic constraints representing regulations, guidelines, policies and laws, are fulfilled by the process. This can be done during both the *process analysis* and the *process monitoring and controlling* phases. During the *process analysis* phase, the process model is statically analyzed to verify that all possible executions comply with the constraints (a-priori compliance checking). During the *process monitoring and controlling* phase, on the other hand, the constraints are either checked against the running process instances (compliance monitoring), or with completed process instances captured in the event logs (a-posteriori compliance checking).

Both *compliance checking* (except for a-priori compliance checking) and *conformance checking* compare the execution of a process against a model, trying to identify deviations between the modeled behavior and the observed one. Additionally, similar approaches are often applied to support both techniques. Thus, many publications treat the terms *compliance checking* and *conformance checking* as synonyms. However, *conformance checking* and *compliance checking* techniques present some differences. *Compliance checking* relies on constraints that describe the admissible behavior of the process only partially. *Conformance checking*, on the other hand, relies on a complete process model, thus describing every possible execution of the process that should be considered admissible. Additionally, *conformance checking* is usually applied only once the execution of a process completes, whereas *compliance checking* can also be applied to running process instances.

2.1.1 Process Monitoring Based on Sensor Data

In the last years, research work with a goal similar to the one of this book, that is, to fully monitor multi-party business processes based on sensor data, was carried out. To compare this work and outline current limitations, we evaluated whether or not the research questions we identified in Section 1.3 were answered. Table 2.1 summarizes the results.

Backmann et al. in [9] extend the BPMN notation with information on how to derive from sensor data streams the activation and termination of activities. Then, they transform this process model into Complex Event Processor (CEP) queries that analyze data streams originated from sensors. This way, it is possible to autonomously detect when activities are executed, thus answering RQ4.1, RQ4.2. Also, violations in the execution order of the activities can be detected without interrupting process monitoring, thus answering RQ2.1 and RQ2.2. As sensor data streams generated by multiple organizations contribute to monitor the process, RQ1.1 and RQ1.2 are also answered.

Table 2.1: Classification of research work on monitoring business processes based on sensor data, based on the research questions defined in Section 1.3. The (+) mark indicates that an answer is provided to the corresponding research question, (-) that the research question was not taken into consideration.

Approach	RQ1.1	RQ1.2	RQ1.3	RQ2.1	RQ2.2	RQ2.3	RQ3.1	RQ3.2	RQ3.3	RQ3.4	RQ4.1	RQ4.2	RQ4.3	RQ4.4
Backmann et al. [9]	+	+	-	+	+	-	-	-	-	-	+	+	-	-
Baumgrass et al. [19,20]	+	+	-	+	-	-	-	-	-	+	+	-	-	-
Mandal et al. [80]	-	-	+	-	-	-	-	+	-	+	+	+	+	-
Gnimpieba et al. [45]	+	+	+	-	-	-	-	-	-	-	+	+	+	-
Stertz et al. [122]	-	-	-	-	-	-	-	-	-	-	+	+	+	-
Senderovich et al. [121]	-	-	-	+	-	-	-	-	-	-	+	+	+	-
Wombacher in [133]	-	-	-	+	-	-	-	+	-	+	+	+	+	-

Baumgrass et al. in [19] and [20] define a software architecture, named UNI-CORN, to standardize and normalize heterogeneous sensor data streams. By executing CEP queries against these streams, UNICORN can identify events related to the activation and termination of activities. In addition, UNICORN supports machine learning techniques to automatically detect unforeseen behaviors that occur during the execution of a process. By coupling UNICORN with a process monitoring platform, named GET Controller, it is then possible to detect when activities are executed, thus answering RQ4.1, RQ4.2, and when the process is incorrectly performed, thus also answering RQ2.1. As sensor data streams generated by multiple organizations are taken into consideration, and monitoring information is distributed among all the organizations participating in the process, RQ1.1 and RQ1.2 are also answered.

Mandal et al. in [80] integrate UNICORN with a case management engine, named CHIMERA. The adoption of case management modeling introduces some degree of flexibility in the monitored processes, thus relieving process designer from modeling all the possible execution scenarios. However, violations in during the execution of a process are not taken into consideration. This work also explicitly takes the IoT into consideration. In particular, data collected by smart objects are used to describe the conditions of the goods participating in the process, which in turn are responsible for determining when activities are executed. This way, RQ1.3, RQ3.4, RQ4.1, RQ4.2, and RQ4.3 are answered. Additionally, the expected lifecycle of the goods, i.e. all the admissible states and transitions that goods are expected to follow when the process is executed, can be modeled. Therefore, RQ3.2 is also addressed. Similarly, Gnimpieba et al. in [45] rely on smart objects to collect sensor data, and rely on a GSM monitoring platform to monitor the execution of multi-party processes, thus addressing RQ1.1, RQ1.2, RQ1.3, RQ4.1, RQ4.2, and RQ4.3.

Stertz et al. in [122] also rely on the IoT, more precisely on Near-field Communication (NFC) tags, to detect when activities are executed based on the position of the physical objects participating in the process. This way, RQ4.1, RQ4.2 and RQ4.3 are addressed. Likewise, Senderovich et al. in [121] rely on the location of physical objects and human operators to detect when activities are executed, thus addressing RQ4.1, RQ4.2 and RQ4.3. In particular, they produce execution logs that are analyzed with standard conformance checking tools to detect deviations from the process model, thus also addressing RQ2.1. Wombacher in [133] monitors the evolution of the physical objects participating in that process and, by comparing it with the information provided by a BPMS, verifies if the execution of the process was performed as expected. This way, RQ2.1, RQ3.2, RQ3.4, RQ4.1, RQ4.2, and RQ4.3 are addressed.

It is worth noting that none of this work provides mechanisms to estimate the impact of process violations. Although [9] states that KPIs can be defined based on the monitoring information, a structured method to define such KPIs is not provided. Even though [22], [80], [45], [122], [121] and [133] take the IoT into consideration, they consider smart objects only as sources of sensor data. Thus, they neither investigate on the possibility of making smart objects self-aware of their conditions and of the process they participate in, nor they tackle the issue of identifying which smart objects are relevant for a specific process execution. Finally, none this work investigates on how to make sure that the data sources are suited to monitor a specific process, which is particularly important when the monitoring platform relies on data provided by external organizations. Therefore, RQ2.3, RQ3.1, RQ3.3, RQ4.4 still remain unanswered.

To answer these research questions, this book adopts and extends methods and techniques related to event data logging, BAM, conformance and compliance checking. Therefore, besides positioning the contributions of this book with respect to the other multi-party monitoring platforms, the state-of-the-art in these research areas is also briefly surveyed in the next subsections.

2.1.2 Event Data Logging

Collecting information about processes being executed is a trivial task when the process is fully automated by a BPMS. Being the BPMS responsible for executing the process, it autonomously decides when activities should be executed. Therefore, the BPMS simply has to annotate whenever it starts or completes an activity. On the other hand, when activities are performed by human operators, the BPMS has to figure out when such activities are actually executed. To do so, a BPMS typically presents a worklist to the human operators [38,112]. Operators then have to select the activities they will perform, and flag them as completed.

However, when no BPMS is present, collecting event data relevant for the process becomes a quite complex task. Since no explicit information on the execution of activities is available, event data have to be derived from the changes the process causes to the environment. Several approaches have been proposed to derive event data from the modifications caused by the process to the operational databases. For example, Gonzalez Lopez de Murillas et al. in [95] analyze the transaction logs generated by the Database Management System (DBMS) hosting the operational database to find operations or tables representing the occurrence of an activity. Nooijen et al. in [102] automatically produce Structured Query Language (SQL) queries to derive event data from the contents of the database. Similarly, Herzberg et al. in [52] extend the syntax of BPMN data objects with constructs to determine, based on the SQL queries executed against a relational database, when activities are started or finished. It is worth noting that, in order to work reliably, such approaches require all the information relevant to the process to be stored in a database. Additionally, they require full read access to the operational databases, which might not be feasible for multi-party processes, where each organization wants to keep its own database private.

Other approaches, such as the one presented by Perez-Castillo et al. in [106], instead of transparently collecting events, modify the software components to introduce logging mechanisms. To do so, they analyze the behavior of the software to determine when events should be recorded in an event log. However, these approaches require both the source code to be available, and the software to be replaceable (which could potentially cause a downtime).

To collect event data for processes that span among different organizations, Engel et al. in [40] rely on Electronic Data Interchange (EDI) messages, widely used in Business-to-Business (B2B) collaborations. Correlations among such messages are identified, thus correlating events belonging to the same process execution. Baumgrass et al [21], on the other hand, rely on CEP to collect event data relevant for the process. More in detail, events coming from sensors, information systems or other applications are filtered and aggregated with CEP queries. This way, it is possible to determine when activities are started or concluded. Di Francescomarino et al. in [43] provide a framework to log events related to the usage of web-based graphical user interfaces. Such events are then used to determine when activities are started or completed, and their dependencies.

2.1.3 Business Activity Monitoring

According to Dahanayake et al. in [34], the main purpose of BAM is to allow organizations to promptly react to issues that occur during the execution of a process. To do so, they analyze events related to the running processes, which can be collected either autonomously or with the help of an event data logging system. By analyzing such events, BAM systems can generate alerts as soon as an issue occurs, and sometimes even before, thus allowing organization to take countermeasures. Additionally, BAM systems can also derive KPIs, thus allowing organizations to have an overall idea on how well the process is performed.

BAM systems can be divided into pure BAM systems and hybrid BAM systems. Pure BAM systems rely solely on rule-based filtering: as soon as one or more events match a specific rule, an alert is produced. This allows organizations to quickly be aware of an issue, as long as a matching rule has been defined (i.e., a shipment took longer than expected). As the alert is produced only after an issue occurred, corrective actions can only compensate the effects of such an issue (i.e., a penalty can be charged to the logistics company, a discount can be offered to the customer).

Hybrid BAM systems, on the other hand, also integrate discovery methods, simulation and reporting tools. This way, it is possible to integrate events with historical information concerning previous executions of the process. This, in turn, makes possible to apply machine learning, constraint satisfaction or Quality of Service (QoS) aggregation approaches, possibly combined, to predict possible issues before they occur, as discussed by Metzger et al. in [89]. Therefore, corrective actions can actually avoid issues to occur.

Initially, BAM focused on properties that were common to all activities to detect or predict the occurrence of an issue. For example, Schmidt and Fleischmann in [119] focus mostly on time metrics, such as the average duration and the number of active instances for each activity. Maamar et al. in [75], on the other hand, focus on the occupation of machines and human operators, and automatically adapt the process when issues are detected. However, techniques that can predicate on proporties that are specific for each single activity have also been proposed. Liu et al. in [71], for example, rely on business artifacts [101] to specify, for each process, which information should be collected when the process is executed. Custom metrics that predicate on such information can then be defined, and KPIs tailored to each specific process can be produced. Kang et al. in [59] and Cabanillas et al. in [24], on the other hand, apply machine learning algorithms to predict the abnormal termination of activities. Even though these approaches are not specific for a certain kind of activity, parameters and support functions that are relevant for the activity to monitor must be chosen for the monitoring to be reliable. For example, Di Ciccio et al. in [36] tailor [24] to detect flight diversions in air transportation activities.

2.1.4 Conformance and Compliance Checking

As previously discussed, conformance checking techniques aim at verifying if a process is executed as expected. As pointed out by Munoz-Gama in [94], conformance checking techniques can be divided into two groups: replay-based and align-based techniques.

Replay-based techniques, as the name suggests, replay event logs to detect if they fit the process model. Whenever a discrepancy is detected, most of these techniques mark the execution as incorrect, and no further analysis on the event logs occur. More sophisticated replay-based techniques, such as the ones presented by Rozinat and van der Aalst in [116] and De Weerdt et al. in [131], continue to replay the event logs, trying to find an execution allowed by the process model that best fits the events.

Align-based techniques such as the ones presented by de Leoni and van der Aalst in [70] and Adriansyah et al. in [3], on the other hand, compare the observed execution captured in the event log to all the admissible executions that are allowed by the model. As long as the observed execution perfectly matches at least one admissible execution, no violation occurred. On the other hand, if no perfect match exists, these techniques try to find the admissible execution mostly similar to the observed one. This way, the impact of a violation on the execution of the process can be assessed, and the activities causing such a violation can be identified.

Both replay-based and align-based conformance checking techniques require the execution of the process to be completed. Therefore, they cannot detect violations between the execution and the model while the process is running. Compliance checking techniques, on the other hand, can deal with violations at runtime. However, their main purpose is different. Instead of verifying that a process is executed as previously defined, compliance checking verifies that normative constraints are not violated when the process is executed. Therefore, compliance checking techniques rely on rules representing constraints that should be fulfilled, rather than on a model representing all and only the executions that are admissible. Nevertheless, most compliance checking techniques can be also adapted to conformance checking, as long as the model is opportunely represented by constraints.

Ly et al. in [73] propose a framework to classify and compare runtime compliance checking techniques. Such a framework consists in ten functionalities that could be offered by compliance monitoring (i.e., runtime compliance checking) techniques.

- **CMF1**. The possibility to express constraints related to time, either qualitatively or quantitatively. For example, requiring that activity *Close container* should follow activity *Fill in container* within 60 minutes. This functionality is fully supported only if both quantitative and qualitative time-related constraints are explicitly supported.
- **CMF2**. The possibility to express constraints related to data. For example, requiring that activity *Take highway* should be executed only if the weight of the container exceeds 5 tons. This functionality is fully supported only if constraints on both activities and data related to the currently monitored process execution are explicitly supported. Also, it must be possible to simultaneously predicate on multiple data objects.
- **CMF3**. The possibility to express constraints related to the organizational resources performing the activities. For example, activities *Fill in container* and *Close container* should be performed by the same person. This functionality is fully supported only if multiple constraints on resources and roles are explicitly supported.
- **CMF4**. The possibility to predicate on the lifecycle of activities. Activities should not be considered as atomic operations, and at least their activation and termination should be considered as distinct. For example, requiring that activity *Close container* should not start until activity *Fill in container* finishes. This functionality is fully supported only if constraints can explicitly predicate on a specific state of an activity (e.g., its activation).
- **CMF5**. The possibility to verify that the lifecycle of an activity is respected. Each activity could have a specific lifecycle, consisting in different states and transitions among these states. For example, requiring that an activity transitioned to *complete* state from *running* state, but not from *inactive* state. This functionality is fully supported only if the constraints on the expected lifecycle of an activity (i.e., which transitions should occur during the execution of the process) can be explicitly defined.

- **CMF6**. The possibility to predicate on multiple instances of the same activity during the same process execution. For example, requiring that activity *Fill in container* completes after all running instances of the activity *Load package* completed. This functionality is fully supported only if multiple executions of the same activity can be detected, distinguishing each execution based on time, data, or resource allocation.
- **CMF7**. The ability to detect and manage violations as soon as they occur. This way, it is possible to take measures to promptly compensate the effects of the violation. This functionality is fully supported only if a violation can be detected when it occurs, the affected portion of the process be clearly identified, and it is possible to continue monitoring the process even after the violation occurred.
- **CMF8**. The ability to detect and manage possible violations before they occur. This way, it is possible to execute the process differently, thus avoiding the violation to occur. This functionality is fully supported only if a violation can be detected before it occurs, or recommendation on how to avoid possible violations are provided to the user.
- **CMF9**. The ability to detect the root cause of a violation. This way, when a violation occurs, it is possible to exactly know which activities, data and/or resources were responsible for such a violation to occur. This functionality is fully supported only if the cause originating a violation can be clearly poited out.
- **CMF10**. The ability to quantify the degree of compliance. This way, besides detecting that one or more constraint were violated it is possible to quantify the impact of such violations on the outcome of the process. This functionality is fully supported only if a metric to estimate the effects of a violation on the process execution is provided.

Once the the framework has been defined, Ly et al. in [73] also survey the state of the art in runtime compliance checking. 14 different techniques are identified and compared with each other by using the introduced framework. In addition, we also applied the framework to the recently emerging predictive process monitoring. Table 2.2 shows the functionalities supported by each of the analyzed techniques.

Santos et al. in [118] apply supervisory control theory to proactively detect compliance violations. Since possible violations are detected before they occur, reactive monitoring and root cause analysis are not considered. Also, this technique does not support data constraints. Similarly, even though not specifically tailored to business processes, ECE rules [23] can be used to monitor compliance. As such, they do not fully support the lifecycle of activities and constraints on data. However, they fully support constraints on time and resources.

BPath [120] is conceived to monitor the execution of workflow-like service compositions by using XPath queries. Therefore, it requires a partial event log to assess the compliance of the running process. It fully support constraints on time, data and resources, but it cannot identify the root cause and the impact of a violation. Dynamo [12,13] also addresses the monitoring of constraints on composite web services. To do so, it observes the messages exchanged by web services. With respect to BPath, constraints on the lifecycle of activities, as well as on the resources can-

Table 2.2: Classification of runtime compliance checking techniques (table adapted from [73]). The (+) mark indicates that the functionality is fully supported, (+/-) that it is partially supported, (-) that it is not supported, and (?) that it was not possible to determine.

Approach	CMF1	CMF2	CMF3	CMF4	CMF5	CMF6	CMF7	CMF8	CMF9	CMF10
Supervisory control theory [118]	+/-	-	+	+	+	-	-	+	-	-
ECE Rules [23]	+	+/-	+	+	-	-	+	-	+/-	+
BPath [120]	+	+	+	+	+/-	+	+	-	-	+/-
Dynamo [12,13]	+	+	+/-	+	?	+	+	-	-	+/-
Gomez et al. [72]	+	-	-	+	?	+/-	+	+	-	-
Giblin et al. [44]	+	?	?	?	?	?	+	?	?	?
Narendra et al. [99]	-	+	+	?	-	+	+	-	-	+
Thullner et al. [125]	+	?	?	?	?	?	+	-	-	?
Mobucon LTL [78,79]	+/-	-	-	+	-	-	+	+	+	+/-
MONPOLY [17]	+	+	+	+/-	+/-	+	+	-	-	-
Halle et al. [51]	+/-	+	+/-	?	?	?	+	?	?	?
Namiri et al. [98]	+/-	+	+	+	-	+	+	-	-	-
MobuconEC [92]	+	+	+	+	+	+	+	-	-	+/-
SeaFlows [74]	+/-	+/-	+/-	+	+	+	+	+	+	+/-
Predictive monitoring [57,77]	+	+	+/-	?	-	?	-	+	-	?

not be defined. On the other hand, mechanisms to partially assess the impact of violations, as well as automatic policies to recover from a violation can be defined.

Gomez et al. in [72] and Giblin et al. in [44] both propose an approach to model and verify constraints on time. The approach presented in [72] relies on constraints programming, whereas the one in [44] adopts timed prepositional temporal logic. In addition, the approach in [72] is also capable of predicting compliance violations. Narendra et al. in [99] propose an approach to selectively verify constraints based on the activities that are currently executed. They rely on first order clauses to model the constraints, which can address data and resources, but not time. Thullner et al. in [125] propose an approach to verify time-based constraints and, when they are violated, to suggest corrective actions.

Mobucon LTL [78,79] relies on constraints modeled with the Declare declarative language. Such constraints are then converted in linear temporal logic expressions, which allow to predicate on time, but not on data and resources. MONPOLY [17] and the approach proposed by Halle et al. in [51] overcome this limitation by both relying on variations of linear temporal logic. This way, constraints on data and resources can also be addressed. Even though not explicitly addressed in MONPOLY, constraints on non-atomic activities and their lifecycle can also be partially modeled.

Namiri et al. in [98] propose an approach to define constraints based on a specific set of control patterns. Constraints are evaluated when a specific event occurs (e.g., an activity is started) and, if the indicated condition is fulfilled, the constraint is violated and the corrective action specified in the rule is executed. Being limited to control patterns, constraints on time can only partially be modeled.

Mobucon EC [92] and SeaFlows [74] describe compliance rules with very powerful languages, Event Calculus and Compliance Rule Graphs respectively, but they do not offer advanced mechanisms to determine the degree of compliance of process instances.

Predictive process monitoring [77], which is implemented by the Nirdizati tool [57], allows to define business goals that can predicate on time, data and, indirectly, resource dependencies. Based on these business goals, it is possible to classify previous execution traces and compare them with the current execution to determine in advance if these goals will be achieved or not, providing a confidence value.

2.2 Declarative Languages

The predominant paradigm for modeling business processes is the one supported by the so-called imperative languages. Widely adopted modeling languages, such as BPMN or Unified Modeling Language (UML) Activity Diagrams, just to name a few, follow this paradigm. The main focus of imperative languages is to model *how* a process should be performed. To do so, when a process is modeled, every possible execution of such a process must be represented in the model. The main disadvantage of this paradigm is its lack of flexibility. In fact, no deviation from the behavior specified in the model is allowed. Consequently, when an imperative model is used for monitoring, whenever a deviation from the expected execution occurs, no further comparison between the model and the execution is possible.

Declarative languages, on the other hand, introduce a complete shift of paradigm. Instead of modeling *how* a process should be performed, they focus on modeling *what* should be done for a process to be executed [112]. With respect to imperative models, that allow only the executions that are explicitly modeled, declarative models allow any execution, as long as it does not contradict the modeled behavior. Therefore, greater variability can be achieved when a process is modeled using a declarative language. In this section, three families of declarative languages, and a language representative for each of them will be presented: constraint-based languages with Declare, artifact-centric languages with GSM, and case management languages with CMMN.

2.2.1 Constraint-based Languages

As the name suggests, constraint-based languages are declarative languages that describe a process in terms of activities that should be performed, and constraints on such activities. Constraints can be divided into mandatory constraints, that must be fulfilled, and optional constraints, that are not required to be fulfilled.

This way, activities are expected to be executed as long as they do not violate any of the mandatory constraints [112].

One of the most widely adopted constraint-based languages is ConDec [107], often referred to as Declare[1] [108]. Process mining solution often rely on ConDec to derive process models from execution traces [28,37,76]. ConDec provides several types of constraints, which can be grouped into three main classes:

- **Existence constraints** specify the cardinality of each activity, i.e., how many times an activity should be executed when the process is performed. This way, it is possible to define an upper and lower bound on the number of times an activity should be executed. For example, during a single process execution, activity A can be required to never be executed, activity B to be executed at least 3 times and at most 5 times, and activity C to be executed exactly 4 times.
- **Relation constraints** specify the dependencies between pairs of activities. This way, it is possible to define that, when an activity is executed, another activity must be executed as well. For example, when activity B is executed, activity D can be forced to be executed immediately after B (alternate response), and B to be preceded by another activity E (precedence).
- **Negation constraints** are the negated version of relation constraints. This way, it is possible to define that, when an activity is executed, another activity must not be executed. For example, when activity C is executed, activity F can be forced to never be executed (negation response), and C not to be immediately preceded by another activity G (negation alternate precedence).

By combining constraints belonging to different classes, ConDec allows to define complex dependencies among activities. However, ConDec does not support conditions on process data. Therefore, the execution of activities cannot be constrained by the state of the resources interacting with the process. For example, a constraint requiring that activity *Ship goods* should be executed only when the container to be shipped is full and closed cannot be modeled with ConDec.

2.2.2 Artifact-centric Languages

Artifact-centric languages are declarative languages that, rather than focusing principally on the activities composing a process, they focus on the business artifacts participating in that process. Business artifacts, firstly introduced by Nigam and Caswell in [101] and also known as business entities, are physical or virtual entities relevant for an organization. Each business artifact (henceforth simply named *artifact*) is characterized by an information model and a lifecycle model. The information model contains all the information relevant to identify and describe the artifact. The lifecycle model, on the other hand, defines all the possible discrete

---

[1] Pesic et al. in [108] refer to Declare as a tool to generate constraint-based language, and to ConDec as one of the languages produced with Declare.

states the artifact may assume, and the admissible transitions from one state to another one. This way, activities composing the process can be seen as the conditions causing the artifact to transition from one state to another one.

One of the most widely known artifact-centric languages is Guard-Stage-Milestone (GSM) [35,53], which will be extensively discussed in Section 4.1. The key elements of the lifecycle model of GSM, as the name suggests, are the following:

- **Stages**. Like activities, stages represent the units of work that are responsible for changing the state of the artifact. Stages can be atomic, thus representing a single task, or can nest other stages.
- **Guards**. They are conditions that determine when a stage are executed.
- **Milestones**. They are conditions that determine when a stage completes its execution. Additionally, milestones can also be used to determine when objectives relevant for the process are achieved. For each milestone, a condition determining when the milestone is no longer valid (invalidation condition) can also be specified.

Conditions associated to guards and milestones can be triggered by both internal and external events. An external event occurs whenever the information model is altered. For example, when a container is shipped, an external event is produced whenever its location changes to alter the information model accordingly. An internal event, on the other hand, is based on the stages that are currently active, and can be used to implicitly define dependencies among stages.

This allows GSM to express quite complex dependencies on both activities and data. However, GSM lacks constructs to define optional dependencies, which would be useful to verify if certain properties are fulfilled when the process is executed. For example, GSM does not allow to specify that, when the container is shipped, the temperature of the container should not exceed 40°C without causing that activity to end.

2.2.3 Case Management Languages

Case management languages can be seen as a specialization of artifact-centric languages that aims at supporting Adaptive Case Management (ACM). With respect to traditional BPM, which focuses on highly structured and repetitive processes, ACM focuses on partially structured, knowledge-intensive processes [100]. To deal with such processes, ACM introduces the concept of case: A case is a set of artifacts, operators, and activities required to achieve a specific goal [123]. A case is firstly modeled with a case management language, in order to define how a generic case should be dealt. When an actual case is handled, the case model is then instantiated.

With respect to traditional process models, operators have more flexibility on the activities composing a case model. In particular, operators can opt not to execute some activities if they are not relevant for that specific case. Additionally, unless explicitly stated when the case is modeled, the execution order of activities can be arbitrarily chosen by the operators. In some occasions, it is even possible to introduce in the case instance activities that were not explicitly defined in the case model.

One of the emerging standards for case modeling is the Case Management Modeling Notation (CMMN) language, an Object Management Group (OMG) standard derived from GSM [67,81]. Compared to GSM, CMMN retains the notion of artifact, which is used to describe all the information relevant for a case. However, it also provides mechanisms to model the goals of the case, the documents produced or required for the case, and the operators that should execute activities. The main constructs of the CMMN language are the following[2]:

- **Stage**. Its semantics is the same as the homonymous construct in GSM, with the exception that a parent stage cannot be closed until all the child stages are closed too. Additionally, stages can be marked as optional.
- **Entry criterion**. Its semantics is the same as the guard construct in GSM.
- **Exit criterion**. It defines under which conditions a stage should be closed.
- **Milestone**. It defines the goals that a case should aim to reach. Compared to the homonymous construct in GSM, it is not responsible for modeling when a stage should close, and is detached from stages. Entry criteria determine when the milestone is achieved.
- **Activity**. With respect to GSM, where a one-to-one association between activities and stages exists, multiple activities can be associated to the same stage. Like stages, entry and exit criteria can be attached to activities. Additionally, activities can be marked as optional, and they can be assigned to a specific class of operators.
- **Case file**. It defines a document that is produced when the case is instantiated, or that is required for the case to be correctly handled and to reach the goals.
- **EventListener**. It allows to explicitly model timers or events raised by the operators that trigger entry or exit criteria.
- **Connector**. It is used to graphically represent the dependencies among the other constructs. For example, to indicate that the completion of an activity triggers the exit criterion of a stage. However, the CMMN specifications do not indicate how and if connectors are derived from entry and exit criteria. As such, connectors are currently used for documentation purposes only, and have no executable semantics.

Even though CMMN has a richer syntax than GSM, and stages and activities can be marked as optional, it still lacks constructs to define optional dependencies among activities. Also, conditions determining if an activity is incorrectly executed cannot be explicitly modeled.

---

[2] For the sake of simplicity, advanced constructs such as decorators, case plans, plan fragments and planning tables are not presented.

2.2.4 Imperative to Declarative Model Translators

As pointed out by Fahland et al. in [42], declarative languages are usually harder to understand than imperative ones, especially when sequential dependencies among activities have to be modeled. Additionally, most business processes have already been modeled using imperative languages. Therefore, it would be desirable to be able to reuse, at least partially, imperative process models when transitioning to declarative languages. By doing so, process designers can start modeling a process with the imperative formalism, which is easy to model and understand. Then, once a declarative model is derived from the imperative one, they can relax some of the constraints to increase flexibility. To this aim, several approaches to translate an imperative model into a declarative one have been proposed.

Köpke and Su in [62, 63] propose an approach to transform a BPMN process model into a GSM equivalent one, enriched with additional stages that group activities according to business goals. The transformation is performed by decomposing the BPMN process model into nested blocks [128], which are transformed into nested stages. Guards and milestones are then derived for each stage according to the type of process block. To improve the understandability of the resulting GSM model, a domain ontology is used to label GSM constructs and to group stages related to the same artifact.

Eshuis and Van Gorp in [41] propose a semi-automated approach to transform a process modeled using UML Activity Diagrams into a GSM model that captures the lifecycle of each involved artifact. Similarly, Kumaran et al. in [66], Meyer and Weske in [90], and Eid-Sabbagh et al. in [39] propose a language-agnostic algorithm to derive the lifecycle of artifacts based on an imperative process model. This is possible as long as each activity has input and output information entities explicitly defined in the model.

2.3 The Internet of Things

The Internet of Things (IoT) is a paradigm that envisions physical objects being uniquely identifiable, being able to interact with the environment, and being capable of communicating with each other and/or with traditional computing devices (i.e. servers, laptop, handheld devices...) thank to the Internet [91]. Key elements of the IoT paradigm are smart objects, physical entities that are uniquely identifiable and addressable, have some computing power, communication interfaces and sensors and/or actuators.

As outlined by Atzori et al. in [8], the IoT is the result of the convergence of different visions: the network-oriented vision, that focuses mostly on the interconnection and communication among smart objects, the things-oriented vision, that focuses on the integration of real world objects inside a common framework, and the semantic-oriented vision, that focuses on analyzing, interpreting and integrating data coming from smart objects.

2.3.1 Enabling technologies of the IoT

Among the key enabling technologies for the IoT, as pointed out in [91] and [8], there are Radio Frequency Identification (RFID) systems, Wireless Sensor Networks (WSNs), and service-oriented middleware.

2.3.1.1 Radio Frequency Identification Systems

RFID systems are asymmetric wireless communication systems made of receivers, named tags, who are only in charge of responding to a radio-frequency message sent by a transmitter, named reader [115].

RFID tags can be both passive (i.e., they require no power source and operate thank to the energy emitted by the reader) or active (i.e., they rely on a power source to operate). They can be read only (i.e., they always respond by sending an unique identifier and other information defined during production) or field-programmable (i.e., data can be written during execution).

As mentioned in [115], RFID systems have been widely employed in ticketing, personal identification, asset tracking, and in the supply chain. For the latter, the EPCglobal standard has been defined for uniquely identifying and retrieving information on RFID tag-equipped products worldwide.

2.3.1.2 Wireless Sensor Networks

A Wireless Sensor Network (WSN), as described in [5] and [109], is a group of multiple small-scale sensing devices responsible for monitoring a specific environment, that forwards sensed values to central node, named sink, via wireless communications.

Advantage of this approach is the decoupling between sensing devices and computational devices: applications can then access sensed values by directly querying sinks, thus ignoring how sensors are deployed. Another important improvement over traditional sensors is the possibility to add, replace and reposition sensing devices without having to reconfigure the infrastructure: sensing devices are designed to deal with changes in the topology by automatically discovering neighbors.

WSNs are widely applied to different application domains, like healthcare, environment monitoring, surveillance and home automation, just to name a few. Their main research issues deal with minimizing power consumption and, consequently, optimizing data transmissions, and increasing computational power.

Being sensing devices battery-powered, minimizing power consumption means servicing or replacing sensing devices less frequently, an activity sometimes difficult if not impossible to carry out. Since most of the power is spent by the communication interface to transmit data to sinks, by optimizing how and when communications occur battery life can be significantly enhanced.

By increasing computational power, on the other hand, it is possible to perform data processing directly at the sensing device level and, by doing so, forwarding only meaningful information (e.g., filtering noisy data or perform data aggregation functions). Since data processing impacts on power consumption, challenges are in minimizing the amount of power and finding a trade-off between the increment of power consumption by more powerful data processing and the reduction of power consumption by avoiding sending unneeded data.

2.3.1.3 Service-Oriented Middleware

A middleware is a software component responsible for standardizing how heterogeneous hardware and/or software architectures communicate with each other by providing a common interface [65]. A Service-oriented middleware is a middleware that exploits Service-Oriented Architecture (SOA) paradigm to allow discovery and composition of hardware/software architectures, which are exposed as services [55].

When dealing with the IoT, as described by Bandyopadhyay et al. in [10], a middleware is responsible for ensuring interoperability among different smart objects both at the syntactic and semantic levels, providing context awareness, supporting discovery and management of smart objects, ensuring security and privacy, and be able to scale.

Concerning interoperability, Coccoli et al. in [31] and Le-Phuoc et al. in [69] propose to adopt the semantic web to standardize data coming from smart objects, thus making exchanged data machine-understandable.

Context awareness is the ability to identify the context into which smart objects operate and use such an information to provide data and functionalities that are relevant for that specific context [104]. Perera et al. in [103] propose to adopt an ontology to classify all the attributes that describe a smart object (e.g., location, sensed data, ...), and use such information to identify which smart objects can deal with a specific context. Da et al. in [135] use a domain specific language to define how smart objects should react to a specific context.

Relatively to the discovery and management of smart objects, Corredor et al. [33] adopts industry standard technologies like web services and UML for describing, retrieving and configuring smart objects. Similarly, Guinard et al. in [48] propose a variation of the web services protocols specifically tailored for the IoT, and a methodology for the discovery and provisioning of services onto smart objects.

Security and privacy are an open issue in the IoT, mainly due to smart objects lacking computational power to deal with security aspects without negatively affecting other operations. Furthermore, as outlined by Chabridon et al. in [27], security and privacy-related aspects conflict with context awareness, since even though fine-grained information are not provided, by combining aggregated data coming from different contexts it is possible to derive such a fine-grained information.

Finally, regarding scalability, Texeira et al. in [124] propose to use approximately-optimal algorithm for service discovery.

2.3.2 Ontologies for the IoT

One of the objectives of the IoT is being able to interact with any of the smart objects. Additionally, smart objects should be able to interact with each other in a M2M fashion. However, one characteristic of the IoT is, like the traditional Internet, the heterogeneity of the systems, protocols and architectures composing it. Therefore, mechanisms to integrate smart objects and middleware produced by different manufacturers are needed.

To ease such an integration, and to formalize the characteristics of smart objects, various approaches propose the usage of ontologies. Guarino et al. in [46] define an ontology as "a formal, explicit specification of a shared conceptualization". As such, an ontology allows to formally describe concepts, their properties, and the relationships among concepts. This way, it is possible to generalize the characteristics of physical and virtual objects, making them understandable by machines. Another advantage of ontologies is the possibility to reference concepts and relationships already defined in other ontologies. By doing so, it is then possible to reuse information that is already standardized and maintained, without having to redo such work.

When applied to the IoT, ontologies can formalize the characteristics of the smart objects and the information collected and processed by them. This way, the integration and communication among smart objects can be automated, even though different technologies are adopted. It is also possible to reference ontologies describing the technologies that realize the smart objects (i.e., the sensors), and ontologies describing the application domain addressed by the smart objects (i.e., manufacturing or logistics).

In the recent years, several ontologies describing the IoT have been proposed. Even though not explicitly targeting the IoT, the W3C SSN ontology [32] provides a well maintained specification of the characteristics of sensors and sensed data. As such, many of the ontologies targeting the IoT adopt and extend SSN.

Kotis et al. in [64] propose an ontology to describe the characteristics of an IoT-based solution. They classify smart objects as sensors, actuators or information processing devices. The ontology also references the SSN ontology to describe the information collected by the sensors. However, this ontology does not explicitly take into consideration composite smart objects, which can have both sensors and actuators. Xu et al. in [134], instead, propose a framework to automatically produce IoT ontologies targeting a specific application domain given specifications written in non-structured text. The resulting ontologies can be easily integrated with the SSN ontology for further specifications.

Cassar et al. in [26] focuses mostly on the services exposed by smart objects. In particular, they propose an ontology to ease the discovery and integration of such IoT-based services. Wang et al. in [129], on the other hand, capture several aspects of a SOA-based smart object. They focus not only on the services exposed, but also on the characteristics of sensors and the sensed data, the location of the smart objects, QoS attributes, and the testing and deployment of the services exposed by smart objects. To do so, they heavily reuse ontologies already covering such aspects, such as SSN and SensorML [6] for the sensors and data. Similarly, Nambi et al.

in [97] propose an ontology to model both SOA-based and physical characteristics of the smart objects.

Cai et al. in [25] propose an ontology to keep track of the characteristics of smart objects throughout their lifecycle. This ontology captures which operations were done during the manufacturing and maintenance of such a smart objects, which parts were used or replaced, and who performed the assembly, testing and maintenance.

Agarwal et al. in [4] present FIESTA-IoT, an ontology to model both the characteristics of both smart objects and the data collected by them. FIESTA-IoT combines the SSN ontology with a subset of the M3 ontology [49] (M3-Lite) – categorizing the functions of the smart objects (e.g., sensor, actuator, etc.), and the sensed information. Other ontologies, such as QU^3 for standardizing the physical properties and the units of measure of sensed data, are also combined. It is also the only ontology to model roaming smart objects, i.e., smart objects that have no fixed location.

Hachem et al. in [50] propose three ontologies supporting the IoT middleware. The device ontology is introduced to describe the physical characteristics of the smart objects. The estimation ontology describes methods and techniques to process and analyze sensed data. Finally, the domain ontology describes physical concepts ad formulas that relate such concepts together (i.e., speed given space and time).

2.3.3 Synergies between the IoT and BPM

While synergies between the IoT and other recent trends, such as Big Data, have been widely established since the beginning, research work on BPM was frequently kept disjoint from the one on the IoT. Nevertheless, the opportunities and challenges that the integration between these two research lines may bring were widely discussed during the last year. As a result, a manifesto of such an integration was recently presented by several academics in the BPM field [56].

According to this manifesto, the advantages of an integration between BPM and the IoT are twofold. Thank to the IoT, BPM can exploit data collected from smart objects to monitor the process and perform business decisions. BPM can also rely on the IoT to let smart objects execute business processes. On the other hand, the IoT can benefit from BPM by relying on business processes to formalize the interaction of different smart objects, thus improving their coordination.

This manifesto also identifies 16 challenges that the integration between these two paradigm may bring, which are the following:

- **IC1 - Placing sensors in a process-oriented way.** Knowing how a business process is structured, and which data are needed by such a process can help to identify the optimum placement for sensors and smart objects.

---

^{3} See http://purl.org/NET/ssnx/qu/qu

- **IC2 - Monitoring manual activities.** Instead of requiring users to manually notify to a BPMS when activities are executed, data collected by smart objects can be used to automate such a task.
- **IC3 - Connecting analytical processes with the IoT.** To make business decisions, reliable and up to date information is required. Smart objects can help to integrate information traditionally available through databases and data warehouses with live data coming from sensors.
- **IC4 - Exploiting the IoT to do process correctness check.** Sensor data collected by smart objects can help to identify issues in a process, such as deadlocks, livelocks or dead activities.
- **IC5 - Dealing with unstructured environments.** Smart objects are often involved in ad-hoc processes, while BPM mostly deals with structured processes, where the structure and the interactions are at least partially known in advance. Therefore, new methodologies and techniques to deal with these ad-hoc processes are required.
- **IC6 - Managing the links between micro processes.** Smart objects are oftentimes involved in micro processes representing habits, instead of complete end-to-end processes. Yet, these micro processes present links and dependencies among each other. Being able to detect and represent such dependencies is then required to have a complete overview.
- **IC7 - Breaking down end-to-end processes.** As mentioned before, smart objects participate in several micro processes that contribute to the execution of complete end-to-end processes. Being able to map portions of these end-to-end processes into micro processes is then required to properly execute them.
- **IC8 - Detecting new processes from data.** Sensor data provided by smart objects can be used to identify processes in a bottom-up fashion. As such, new processes can be discovered, and a certain degree of freedom in their execution can be introduced. However, constraints on such discovered processes should be introduced to optimize the usage of resources, and to prioritize specific goals.
- **IC9 - Specifying the autonomy level of smart objects.** In the IoT, smart objects autonomously react to events by executing tasks or process portions. While a certain level of autonomy is desirable, mechanisms to override such an autonomy in favor of a centralized supervision, or to veto certain actions should be introduced.
- **IC10 - Specifying the "social" roles of smart objects.** Since the goals of an organization's process may differ from the ones defined for the smart objects participating in such a process, governance mechanisms to resolve these conflicts are required.
- **IC11 - Concretizing abstract process models.** Oftentimes, business processes are firstly modeled in an abstract way, in order to capture the general behavior of the process, then dynamically turned into an executable model. To do so in the IoT context, matchmaking mechanisms that map activities or process portions of the abstract model to the smart objects according to their capabilities are required.
- **IC12 - Dealing with new situations.** To manage unplanned situations, BPM technologies such as task recommendation or the conditioned execution

of processes and activities can be adopted. To this aim, sensor data collected by smart objects can provide more accurate information to contextualize such situations, and to compare them against previous ones.

- **IC13 - Bridging the gap between process-based and event-based systems.** The IoT generates a large amount of heterogeneous sensor data. Being able to infer complex events from such data, and to correlate these events to process instances is far from trivial. To this aim, process mining techniques from one hand, and Complex Event Processor (CEP) from the other hand, can be helpful to bridge such a gap.
- **IC14 - Improving online conformance checking.** As previously stated, sensor data collected by smart objects can be used to determine when activities are executed. This information can then be used to detect discrepancies between the process being executed and the planned one as soon as they occur.
- **IC15 - Improving resource utilization optimization.** In a pure IoT paradigm, the capabilities of smart objects are dimensioned according to the expected situations they have to react to. Exploiting BPM techniques, on the other hand, the capabilities can be optimized globally, with respect to the processes interacting with such smart objects.
- **IC16 - Improving resource monitoring and quality of task execution.** The IoT can help to identify issues in the utilization of resources participating in the process, such as excessive stress in human operators, or faults in machines.

These challenges are also useful to frame existing research work. Table 2.3 shows the relationships among these challenges and some of the approaches dealing with the integration of IoT with BPM that were presented in the last years, excluding artifact-driven process monitoring, which is discussed in this book, and will be compared with these approaches in Chapter 8. As the focus of this book is on monitoring business processes, challenges IC3, IC8, IC9, IC10, IC11, and IC15 are not considered. The criteria that were followed to evaluate to which extent these solutions address these challenges are reported in Appendix B.

Stertz et al. in [122] rely on NFC tags and readers, a derivative of RFID systems, to detect when activities are executed. Whenever a human operator performs a certain activity, it simply puts the physical objects involved in such an activity, which are equipped with an NFC tag, near an NFC reader. This way, it is possible to know when an activity was started and by whom. Since a traditional BPMS is adopted, activities are expected to be executed according to an imperative process model. Additionally, the process model is used to determine which objects are involved with each activity, and which is the appropriate location to place the NFC scanners. Gnimpieba et al. in [45] also adopt a similar approach: they rely on sensor data coming from the physical objects participating in the process to determine when activities are executed. However, they rely on a monitoring engine using the GSM artifact-driven notation. In particular, they use GSM to model the conditions, based on sensor data, that determine the activation and termination of each activity. This allows to define more flexible process models than the ones that a traditional BPMS supports, thus introducing more flexibility. Senderovich et al. in [121] also rely on the location of physical objects and human operators to detect when activities are executed. In particular, they compare the position of all the

Table 2.3: Classification (inspired by [56]) of approaches dealing with the integration of IoT with BPM. The (+) mark indicates that the functionality is fully supported, (+/i) that it is partially supported, (-) that it is not supported.

Research work	IC1	IC2	IC4	IC5	IC6	IC7	IC12	IC13	IC14	IC16
Stertz et al. [122]	+/-	+	-	-	-	-	-	-	+/-	-
Mandal et al. [80]	-	+/-	-	+/-	+/-	+	+/-	+	+/-	-
Senderovich et al. [121]	-	+	-	-	-	-	+/-	+	+/-	+/-
Wombacher [133]	-	-	+	-	-	-	+/-	-	+	-
Weber et al. [130]	-	+/-	-	-	-	-	-	-	-	+
Gnimpieba et al. [45]	-	+	-	+	+/-	+/-	+/-	+	+/-	-
Knoch et al. [61]	+	+	+	-	+	-	-	+/-	+/-	+

objects and operators required for an activity to start with the location where such an activity should be executed. However, instead of relying on a BPMS to monitor the process being executed, they directly generate an execution log, which can then be analyzed with standard conformance checking tools to detect deviations from the process model.

Wombacher in [133] performs online conformance checking by comparing the evolution of a process model, as recorded in a BPMS, with the evolution of the physical objects participating in that process. Whenever an activity is marked as executed by the BPMS, but the results of such an execution are not reflected in the conditions of the physical objects, the presence of a potential issue is recorded. Similarly, when a change in the condition of a physical object is detected while no activity related to such an object was executed, an alert is raised. Such information are then analyzed to determine if these issues are caused by the process not being executed as expected, sensors being faulty, or the process model being incorrect. Knoch et al. in [61], on the other hand, rely on information coming from various sensors, such as infrared cameras and ultrasonic sensors, to infer when activities are executed. This information is then used to determine performance metrics, such as the usage of resources and the workload required to execute an activities. Similarly, Weber et al. in [130] rely on neurophysiological measurements, such as eye tracking, to determine the cognitive load of human operators when executing a task. This information can be used to both infer which activities are being performed, and the impact of the workload on the operator.

Mandal et al. in [80] integrate a case management engine with a CEP. This way, sensor data coming from smart objects can be processed and aggregated into complex events, which can then trigger actions in the case management engine. The adoption of a case management engine instead of a BPMS allows to dynamically compose process fragments into a full end-to-end business process, thus increasing flexibility over unplanned situations.

Chapter 3
Artifact-driven Process Monitoring Overview

As anticipated in Chapter 1, one of the main issues in monitoring business processes is identifying when activities are performed. As long as a process can be fully automated and is confined inside a single organization, a traditional BPMS can easily solve this issue. In fact, a BPMS can autonomously keep track of any activity it automates. Such activities typically consist in the invocation of external programs or web services, and, to some extent, form-based operations (i.e., filling in a tax declaration form). Being responsible for both the execution and the monitoring of those activities, a BPMS only has to log whenever it starts or completes the execution of those activities.

However, when activities are not automated, a BPMS cannot directly know when they start or end. This occurs whenever an activity is performed by a human being, and no interaction with the BPMS is required to actually perform the activity (i.e., inspecting a package). In this case, the only way for a BPMS to know when a non-automated activity is performed would be to receive external notifications. To do so, a BPMS typically proposes, via user interface, a list of activities to the operator responsible for their execution. It is then up to the operator to notify the BPMS when each activity is initiated and when it completes. However, this task requires the user to stop its own duties to interact with the BPMS. Therefore, it is prone to being forgotten, delayed, or erratically performed, either accidentally or intentionally.

This issue traditionally characterizes human-centric business processes, where most of the activities are not automated. However, when it comes to multi-party processes, the issue becomes critical. Firstly, it is more difficult to force users outside the boundaries of an organization to send notifications. Secondly, if each organizations adopts its own BPMS, and those BPMSs are not federated, explicit notifications are required even for automated activities, if they are performed by the other organizations.

To address this issue, we propose to exploit the IoT paradigm to perform process monitoring directly on the physical objects embodying the artifacts interacting with the process. Thank to this novel approach, named artifact-driven process monitoring, the monitoring infrastructure is no longer confined on the premises of a single organization. Instead, process monitoring can be performed in close

G. Meroni: Artifact-Driven Business Process Monitoring, LNBIP 368, pp. 37–44, 2019
https://doi.org/10.1007/978-3-030-32412-4_3

contact with the operators responsible for executing business activities, even if these activities are carried out by another organization. Additionally, smart objects can autonomously infer when activities are executed and if the artifacts they embody are properly manipulated.

3.1 Motivating Example

To better understand the issues in monitoring multi-party processes involving manual activities, and how such issues are addressed with our artifact-driven process monitoring approach, a real scenario taken from the logistics domain is presented. Such a scenario was discussed and validated in the context of the ITS Italy 2020 research project, and it will be used throughout this book. However, logistics is only one of the possible application domains that will benefit from artifact-driven process monitoring. In fact, as long as physical objects interact with a process, and such objects can be turned into smart objects, artifact-driven process monitoring can be beneficial to every business process that is not automated and/or spun among multiple organizations.

A Piedmontese manufacturer has to send its goods, which are sensitive to high temperatures, to one of its Lombard customers. To do so, it organizes a so-called three-leg multi-modal shipment, which is shown in Figure 3.1: Firstly, a shipping container containing the goods is sent to an inland terminal located near Turin by truck (first leg, the so-called first mile). Then, the container is sent to another inland terminal located near Milan by train (second leg). Finally, the container is delivered to the customer by truck (third leg, the so-called last mile). To perform such a shipment, it outsources each leg to a different carrier.

Figure 3.2 shows the BPMN representation of the first mile. Firstly, the manufacturer fills a shipping container with the goods to be shipped. At the same time, the carrier drives one of its trucks to the manufacturer's premises. Once the truck arrives, the manufacturer attaches the container to the truck. Then, the truck starts driving to the inland terminal: it travels on the highway to Turin until either it is midnight or the truck reaches the inland terminal's premises. In the first case, a break is taken and, once the break ends, the truck starts traveling again. In the second case, one operator at the inland terminal inspects the goods, then detaches the container from the truck. If, when the truck is driving to the inland terminal's premises, the container overheats, the process has to be aborted.

A peculiarity of this process is that no organization fully controls it. The manufacturer is responsible only for filling in the container and attaching it to the carrier's truck. The carrier can only reach the manufacturer and the inland terminal, or take a break. Finally, the operator at the inland terminal can only either inspect the goods or detach the container.

Another peculiarity of this process is that some physical objects belonging to an organization are manipulated by the other organizations as well. Remarkably, the container, which belongs to the manufacturer, is handled by the carrier when it is delivered to the inland terminal, and by the operator at the inland terminal

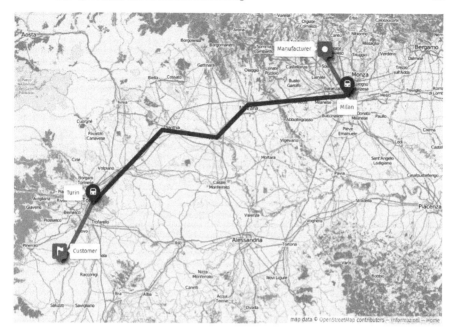

Fig. 3.1: Route followed by the three-leg multi-modal shipment. This map was produced with umap (`http://umap.openstreetmap.fr/`).

when it is inspected and detached. Also the truck, which belongs to the carrier, is manipulated by the manufacturer when the container is attached, and by the operator at the inland terminal when the container is detached.

Finally, some of the activities composing this process require human intervention to be performed. While the container could potentially be filled in, attached to and detached from the truck automatically, a human operator would still be required to drive the truck, and to inspect the goods inside the container.

3.2 Introducing Artifact-driven Process Monitoring

The IoT paradigm aims at providing sensing capabilities, computational power, and transmission interfaces to physical objects, making them *smart*. This way, these smart objects can collect information on the environment they operate, as well as on their physical conditions. Additionally, they can exchange such an information with humans, information systems, and other smart objects.

Thank to the capabilities that smart objects offer, they can be made aware of the process they operate. By doing so, the process can be then directly monitored by the artifacts, embodied by the smart objects, participating in that process.

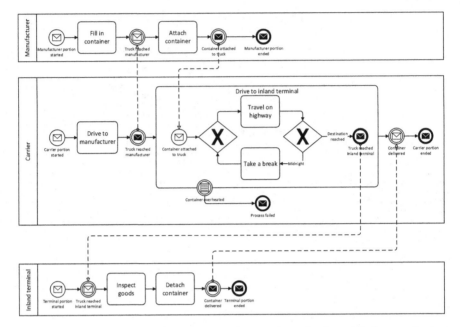

Fig. 3.2: BPMN collaboration diagram showing how the first mile is organized.

We name such an approach artifact-driven process monitoring. This brings several advatages:

- Firstly, sensors can be used to detect when the state of the artifacts embodied by the smart objects changes. By comparing such a change of state with the expected lifecycle of the artifact, smart objects can then detect if the artifact is misused, and raise an alert. For example, an accelerometer inside the container could be used to determine if the container is dropped.
- Additionally, smart objects can notify changes of their state to the other smart objects participating in the process. As extensively discussed in Chapter 5, changes in the state of the artifacts can also be used to infer when activities are executed. This way, smart objects can also determine when activities are executed. For example, detecting that an empty container is opened could mean that the manufacturer started filling in the container.
- Finally, as the whole monitoring infrastructure is enclosed in a smart object, which can easily cross the boundary of an organization, monitoring a multi-party business process becomes easier. In fact, a smart object can stay in close contact with the process to be monitored. Therefore, it can autonomously collect all the information relevant for the process, without having to be federated with the information systems of the external organizations. For example, based on the position of the hook of a truck, the carrier could know when a container

is attached or detached to the truck, even though these activities are performed by the other organizations.

When applied to the process presented in Section 3.1, artifact-driven process monitoring would allow the manufacturer to know if the container is incorrectly manipulated even after it leaves the manufacturer's premises. Also, it would allow the manufacturer to monitor when the activities carried out by the carrier and the operator at the inland terminal are performed, and if they comply with the control flow defined in the process model (i.e., if the goods are not inspected at the inland terminal's premises).

Currently, the applicability of this artifact-driven monitoring approach is limited by the cost and size of the physical objects. For example, it makes sense to turn trucks or shipping containers into smart objects, since they are quite large, heavy and expensive objects per-se. On the other hand, it may not be economically feasible to equip the packaging of the goods with a computing device powerful enough to run the monitoring platform. However, we believe that, thank to the continuously increasing computational power, and the constant reduction both in size and cost of single board computing devices, in the next years more and more physical objects will be suited for artifact-driven monitoring.

3.3 Reference Architecture

To be practically implemented, an artifact-driven process monitoring platform requires a reference software architecture to be defined. During the design of this reference architecture, we identified the following architectural requirements:

- **AR1:** The platform should be modular. This way, modifications that affect only a portion of the platform should not require the whole platform to be redeployed.
- **AR2:** The platform should allow smart objects to exchange information on their states. This way, it is possible to predicate on the state of multiple smart objects to identify when each activity is run.
- **AR3:** The platform should be hardware-agnostic as much as possible. This way, the effort required to tailor an artifact-driven monitoring platform to a smart object is minimized.

The result of this design phase is the reference architecture shown in Figure 3.3, which is organized along four main modules, each one representing an incremental step to achieve artifact-driven process monitoring:

- **On-board Sensors Gateway.** This module runs on each smart object, and is responsible for periodically collecting the values coming from the sensors attached to the smart object, and transforming them into messages describing the physical properties of the object[1]. For example, if a truck is equipped with a

---

[1] Sampling time and type of interaction (pull/push) can be configured given the types of the attached sensors.

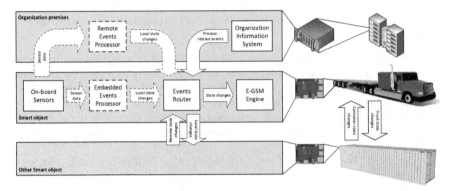

Fig. 3.3: Reference architecture of our artifact-driven monitoring platform.

GPS receiver and a speedometer, the On-board Sensors Gateway periodically collects data from these sensors, and transforms them into messages on the speed and the location of the truck.

- **Events Processor**. This module takes in input the messages produced by the On-board Sensors Gateway, and infers from the physical properties of the object the discrete state of the artifact embodied by such an object. For example, the Events Processor of the truck can determine if the truck is moving based on the speed.

  In the reference architecture, the technology behind this module is intentionally left unspecified: This module can be implemented with a rule engine, a CEP, or ad-hoc code. Additionally, based on the computational power required to derive the state of the artifact and on the capabilities of the smart object, this module can be deployed either on a smart object, or at the organization's premises (and then remotely accessed by the smart object).

- **Events Router**. This module, running on each smart object, is responsible for transmitting the current state of the artifact embodied by the smart object to the other smart objects participating in the same process instance. Additionally, it receives information on the state of the other artifacts from the other smart objects, and events related to the process instance but unrelated to the artifacts from the information system of the organization owning that smart object. For example, the Events Router of the truck transmits changes in the state of the truck to the container, and receives information on the state of the container from that smart object. Also, it receives notifications on when the process is started or concluded from the information system of the carrier.

  For the Events Router to work properly, the identity of the other smart objects participating in the same process instance must be specified. This will be discussed in detail in Chapter 5.

- **E-GSM Engine**. This module, running on each smart object, is responsible for determining, based on the state of the artifacts participating in the process, when activities are executed. It also determines if the process execution deviates from the one previously agreed between the organizations, and opportunely flag the portions of the process affected by such a deviation. To do so, the E-GSM Engine requires the process monitored by the smart object to be represented using the E-GSM notation. The syntax of the E-GSM notation, as well as the motivations behind the choice of such a formalism to monitor the process will be extensively discussed in Chapter 4.

Since the output of each module is used as an input for the subsequent one (i.e., the output of the On-board Sensors Gateway is used as input for the Events Processor), we designed the architecture following the pipe-and-filter design pattern. This way, each module depends on the other ones only in terms of inputs and outputs. Therefore, as long as the input and output formats remain the same, each module can be replaced without requiring the other ones to be altered, thus achieving AR1.

Additionally, the pipe-and-filter pattern simplifies the exchange of information among smart objects, thus achieving AR2. In fact, for smart objects to be aware of the state of the other artifacts, their Events Router modules simply have to forward the messages received from the Events Processor module.

Finally, this modular architecture allows to easily adapt the platform to several smart objects. In particular, the hardware-dependent portions are confined inside the On-board Sensors Gateway module. Consequently, when the platform has to be deployed on a new type of smart object, only this module has to be altered, thus minimizing the deployment efforts and achieving AR3.

3.4 Summary

This chapter discussed how the IoT can be helpful to monitor non-automated, inter-organizational business processes. To better explain the main issues when dealing with such a kind of processes, a motivating example from the logistics domain has been introduced. The aspects that make such a process difficult to monitor have then been discussed.

Then, the advantages of adopting the IoT paradigm to move the monitoring tasks directly onto the physical objects participating in the process have been discussed. In particular, it has been pointed out that physical objects can be in close contact with the activities composing process, and can travel across multiple organizations.

Finally, a reference architecture supporting the artifact-driven monitoring approach has been presented. The modules that run on the physical objects participating in a process have been briefly introduced. Additionally, the role of such modules in determining the state of the artifacts embodied by the physical objects and the overall execution of the process has been discussed.

Chapter 4
E-GSM: an Artifact-centric Language for Process Monitoring

One important issue in process monitoring is being able to promptly identify when the process is not executed as expected. As pointed out in Section 2.1.3, BAM techniques primarily focus on providing performance metrics (e.g., response time, throughput, etc.) and on determining if activities or process portions will be incorrectly executed. However, when it comes to determining if the execution order of activities respects the one defined in the process model, or if the artifacts are managed as expected, the issue is still not solved. Conformance checking techniques, which are used to detect discrepancies between the process model and the executions, are meant to be applied after the process finished. As such, they cannot provide timely notifications when the process is executed. Runtime compliance checking techniques, on the other hand, are only meant to verify if the execution complies with constraints representing regulatory requirements and policies. As such, compliance constraints represent only a small subset of the behavior defined in the process model. Even though the process model could potentially be expressed with compliance constraints, the complexity of the resulting model would be quite high.

To address this issue, artifact-driven process monitoring relies on an extension of the Guard-Stage-Milestone (GSM) declarative language, named E-GSM, to represent the process to be monitored. In declarative languages the dependencies among activities are treated as descriptive rather than prescriptive. Consequently, declarative languages can be used to keep track of process executions that do not exactly follow the expected execution flow.

4.1 The Guard-Stage-Milestone Artifact-centric Language

To identify the language that allows to autonomously and continuously monitor a business process, we defined the following requirements:

- **LR1:** The execution semantics of the language should distinguish between activities that are being executed, and activities that are either completed or not yet executed.

© Springer Nature Switzerland AG 2019
G. Meroni: Artifact-Driven Business Process Monitoring, LNBIP 368, pp. 45–60, 2019
https://doi.org/10.1007/978-3-030-32412-4_4

- **LR2:** The language should provide constructs to determine, based on external data and events, when activities are being executed.
- **LR3:** The language should allow both the activities composing the process and the lifecycle of the artifacts participating in the process to be modeled.
- **LR4:** The language should not enforce the execution of the process to strictly adhere to the dependencies among activities defined in the model.
- **LR5:** The language should provide constructs to determine, based on external data and events, when an activity fails to be successfully completed. Failures in the execution of an activity should not depend on dependency constraints.

Among the existing declarative languages, we found GSM to be the best candidate for monitoring business processes and detecting violations at runtime. As anticipated in Section 2.2.2, GSM belongs to the artifact-centric language family. As such, it represents a process in terms of the artifacts that participate in that process. For each artifact, GSM captures its characteristics, and defines the activities responsible for altering them. To do so, GSM relies on four main constructs, which are shown in Figure 4.1:

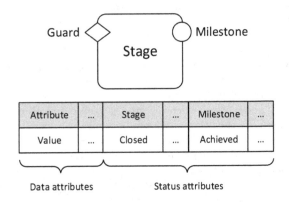

Fig. 4.1: GSM graphical representation.

- **Stage.** This construct represents the activities responsible for interacting with and altering the artifact. A stage can be atomic, thus representing a single activity, or can nest other stages hierarchically.
- **Guard.** This construct represents the condition that causes an activity to be executed. As such, each stage must have at least one guard, and each guard must be associated to exactly one stage.
- **Milestone.** This construct represents the condition that causes an activity to stop being executed. As such, each stage must have at least one milestone, and each milestone must be associated to exactly one stage. For each milestone, it is also possible to define a condition that makes it no longer valid, thus causing the associated stage to be executed again.

- **Information model.** This construct represents the current state of the artifact. As such, it is composed of two parts: data attributes, that describe the characteristics of the artifact, and status attributes, that keep track of the stages being executed and the milestones achieved. When the GSM model is instantiated, the set of values assumed by the attributes in the information model at a certain point in time is called snapshot.

Each stage can be either *open* or *closed*. When a stage is *closed*, all the nested stages, if any, are *closed* as well, and no activity can be executed. When a stage is *open*, the nested stages, if any, can be either opened or closed, and the activities can be executed. Initially, a stage is *closed*. When one of the guards of a *closed* stage is activated, and the parent stage is *open*, that stage becomes *opened*. When one of the milestones of an *open* stage is achieved, the stage becomes *closed*. Also, when one of the milestones of a *closed* stage is invalidated, the stage becomes *open* again.

To represent the conditions on guards and milestones, GSM adopts Event-Condition-Action (ECA) rules. An ECA rule is an [on e] [if c] expression, that is triggered when an event e occurs and the condition c is true. When [on e] is missing, the ECA is triggered once c becomes true, when [if c] is missing, the ECA is triggered once e occurs. An event e can be either internal or external. An internal event occurs when a stage transitions from open to closed or vice versa, or when a milestone is achieved or invalidated. An external event, on the other hand, occurs when the data attributes in the information model are altered.

Whenever an external event occurs, it alters the data attributes in the information model, and causes the ECA rules that are triggered by that event to be evaluated. If one of these ECA rules is satisfied, it alters the status attributes in the information model. This causes other ECA rules, predicating on that portion of the information model, to be evaluated. If one of these ECA rules is satisfied, another change in the status attributes occurs. The alternation of changes in the information model, followed by ECA rules being evaluated, continues until no ECA rule is satisfied and, consequently, no further change in the information model occurs. A change in exactly one attribute in the information model is called micro-step. On the other hand, the sequence of changes that cause the information model to transition from the snapshot before the external event occurs to the snapshot where no further change occurs is called b-step. As such, one b-step consists in one or more micro-steps.

To avoid an external event to cause an endless loop, i.e., an infinite sequence of micro-steps, a b-step is allowed to change the value of each status value at most once (i.e., toggle-once). As such, when an external event occurs, each milestone can transition from achieved to invalidated, or vice versa, at most once. Similarly, each stage can transition from open to closed, or vice versa, at most once.

The main advantages of GSM with respect to the other languages are the following:

- GSM provides constructs to keep track of the activities being executed, regardless of their complexity. Stages can be used to model atomic activities, process portions, and subprocesses. Since stages can be nested, they can be used to model which atomic activities compose a process portion or a subprocess. As running stages are marked as open, and stages not jet run or whose execution completed are marked as closed, they can be used to keep track of which activities are executed, thus fulfilling LR1.

- GSM provides constructs to determine from external data when activities are executed. Thanks to guards and milestones, it is possible to define expressions to detect, based on the data collected by the smart objects impersonating the artifacts, when activities are started or ended (i.e., when stages are opened or closed). Therefore, LR2 is fulfilled.
- GSM is an artifact-centric language. This allows to model the process from the point of view of the artifacts. Consequently, the process is not only defined as a set of activities and dependencies among them, but also as a set of states that an artifact is supposed to assume, and transitions from one state to another one. This way, GSM can keep track of both the activities being executed, and the states that artifacts assume, thus fulfilling LR3.

When assessing the various declarative languages, CMMN was also considered for process monitoring. Indeed, CMMN also provides constructs, namely Exit criteria and **Exit criteria**, to define from external data when **Stages** are executed. However, in CMMN **Stages** are only used to represent process portions, whereas **Activities** are used to model atomic tasks. Therefore, multiple **Activities** can be contained inside a single **Stage**. Additionally, **Entry** and **Exit criteria** cannot be associated to **Activities**. Therefore, to fulfill LR1 and LR2, the CMMN standard should be subverted by either imposing a one-to-one relationship between atomic **Stages** and **Activities**, or by allowing **Entry** and **Exit criteria** to be also associated to activities. An other issue in the adoption of CMMN is its tight relationship with cases. Therefore, using CMMN to model the lifecycle of the artifacts participating in the process, which is required to fulfill LR3, would subvert the purpose of the language.

4.2 Extending GSM

Despite the previously discussed capabilities of GSM that can help monitoring a process, GSM cannot be used for monitoring as-is. In particular, it lacks constructs to explicitly define the *expected* execution flow, which are required to fulfill LR4: even though guards can be used to model the dependencies among stages, they will cause the associated stage to be opened as soon as the dependency is met. This is not desirable when the process is monitored, since the activity represented by that stage could not be executed even if the dependency is met. Additionally, GSM also lacks constructs to define when activities represented by stages are incorrectly performed, which are required to fulfill LR5.

To address these issues, we propose an extension of GSM, named E-GSM, to fully support process monitoring. As shown in Figure 4.2, E-GSM comprises the following constructs:

- **Stage**. It is identical to the homonymous GSM construct.

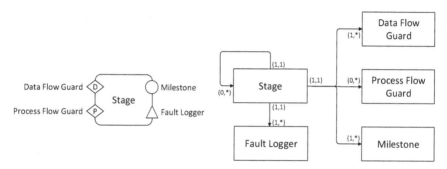

Fig. 4.2: E-GSM graphical representation (left) and metamodel (right). The information model is not depicted.

- **Data flow guard.** It is an ECA rule[1] that, if true, causes the associated stage to be opened. It replaces the GSM guard construct.
- **Milestone.** It is identical to the homonymous GSM construct.
- **Process flow guard.** It is a boolean expression that models the expected execution flow, thus addressing LR4. To do so, it predicates on the activation of the data flow guards and the achievement of the milestones. The expression is evaluated when at least one of the data flow guards associated to the same stage as the process flow guard is triggered, and before the stage becomes opened. If the expression is true, the stage complies with the expected execution, otherwise the stage has been activated without respecting the expected execution flow.
- **Fault logger.** It is an ECA rule that, if true, indicates that the associated stage is faulty (i.e., something went wrong when the activity represented by that stage was executed), thus addressing LR5. A faulty stage does not imply its termination, as the termination is only determined by the achievement of a milestone.
- **Information model.** It is identical to the homonymous GSM construct.

As anticipated in the previous section, the lifecycle of a GSM stage is composed of only two states: open, if the activity represented by that stage is being executed, and closed, if the activity finished its execution, or it has never been executed. On the other hand, the lifecycle of an E-GSM stage is composed of several states and, as shown in Figure 4.3, it is organized along three orthogonal perspectives: status, outcome, and compliance[2]. It is worth noting that E-GSM does not alter the operational semantics of GSM, but extends it with new constructs and perspectives.

---

[1] An ECA rule is an [on e] [if c] expression, that is triggered when an event e occurs and the condition c is true. When [on e] is missing, the ECA rule is triggered once c becomes true, when [if c] is missing, the ECA rule is triggered once e occurs.

[2] In this paper we use the the the notation introduced in [53], so we write $S.DFG_i$, $S.PFG_k$, $S.FL_1$ to indicate the activation of a data flow guard, process flow guard, or a fault logger associated with stage S, $+S.M_j$ ($-S.M_j$) to indicate the achievement (invalidation) of a milestone M_j, $S.M_j$ to indicate that stage S is closed and a milestone M_j is achieved, and Active(S) to indicate that stage S is opened.

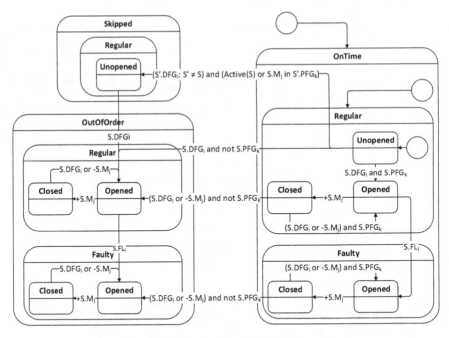

Fig. 4.3: Lifecycle of an E-GSM stage S.

- The **execution status** captures the status of a stage: *unopened, opened* or *closed*. A stage is *unopened* if its data flow guards have never been triggered. A stage can become *opened* only if it is *unopened* or *closed* and the parent stage is *opened*. In addition, at least one of its data flow guards must be triggered ($S.DFG_i$). A stage becomes *closed* if it is *opened* and a milestone is achieved ($+S.M_j$), or if the parent stage becomes *closed*.
- The **execution outcome** captures the situation of a stage. A stage is declared *regular* by default. When the stage is *opened* and one of its fault loggers is triggered ($S.FL_1$), it becomes *faulty*.
- The **execution compliance** captures the compliance of each stage with the expected execution flow. A stage is declared *onTime* by default. It can become *outOfOrder* (according to the expected execution flow) when one of its data flow guards is triggered but none of its process flow guards holds ($S.DFG_i$ and $not(S.PFG_k)$). If a stage S' is declared *outOfOrder*, every other *onTime* stage S that precedes S' ($S.M_j$ or $Active(S) \in S'.PFG_k$) is declared *skipped*. If a stage is *skipped*, once one of its data flow guards is triggered ($S.DFG_i$), it becomes *outOfOrder*.

The combination of these three perspectives says that the whole lifecycle assumes that a stage is initially *onTime, regular,* and *unopened.* data flow guards drive the change of execution status. fault loggers drive the outcome, while process flow guards are in charge of the compliance. With respect to GSM, E-GSM interprets reopening a *closed* stage as a new iteration of that process portion. Therefore, once a parent stage is reopened (i.e., it moves from *closed* to *opened*), the lifecycle of all its child stages will restart from scratch.

To give an idea on how a process can be directly modeled in E-GSM, the following example is introduced. To deliver its goods to one of its customers, located near Naples, a Tuscan manufacturer relies on the three-legged multi-modal shipment shown in Figure 4.4. Firstly, a shipping container containing the goods is sent to the port of Livorno by truck. Once there, the container is sent to the port of Naples.Normally, sea shipment is adopted for this leg. However, during the month of December, the railroad connection between the two ports is adopted instead. Finally, the container is delivered to the customer by truck. During the whole delivery process, the container should never overheat.

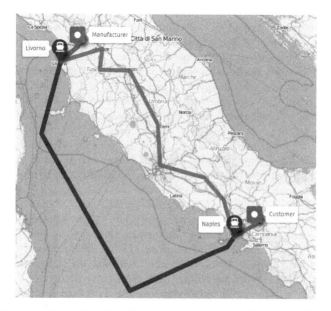

Fig. 4.4: Route followed by the three-leg multi-modal shipment. This map was produced with umap (http://umap.openstreetmap.fr/).

Figure 4.5 shows the E-GSM model of this process. Since this process consists in four activities, four stages are introduced: *ShipToLivorno*, *ShipToNaplesByRail*, *ShipToNaplesBySea*, and *DeliverToCustomer*. To be executed, all these activities require a mean of transport (e.g., a truck), and the container. Thus, to determine when these activities are executed, Data Flow Guards and Milestones associated to the corresponding stages can predicate on the conditions of these two artifacts. In turn, these conditions can be represented as a set of discrete states that the artifacts can assume. For example, the following ECA rule is defined for *ShipToLivorno.DFG*1: `on container_e or truck_e if container[hooked]` and truck[manufacturer]. This way, $ShipToLivorno.DFG1$ is triggered when either the container or the truck assumes a new state, and it is fired only if the container is attached to the truck, and the truck is located at the manufacturer's premises. Similarly, $ShipToLivorno.M1$ is triggered when either the container or the truck leaves the current state, and it is fired only if the container is detached to the truck, and the truck is located at the port of Livorno (on container_1 or truck_1 if container[unhooked] and truck[livorno]).

When an activity has never been executed, none of the Milestones of the corresponding stage is already achieved. Otherwise, that would mean that the activity was already executed at least once. Thus, to indicate that each leg should be executed only once during the shipment process, the Process Flow Guard associated to the corresponding stage requires that none of the Milestones of that stage be achieved. Thus, the boolean expression not ShipToLivorno.M1 is defined for $ShipToLivorno.PFG1$.

To indicate that $ShipToNaplesByRail$ and $ShipToNaplesBySea$ are mutually exclusive, and that $ShipToNaplesByRail$ should be executed only during the month of December, their Process Flow Guard predicates also on the current month. Therefore, the boolean expressions not ShipToNaplesByRail.M1 and month[december] and not ShipToNaplesBySea.M1 and not month[december] are defined for $ShipToNaplesByRail.M1$ and $ShipToNaplesBySea.M1$, instead.

Additionally, a new stage Exc, that encloses $ShipToNaplesByRail$ and $ShipToNaplesBySea$, is introduced. This way, Exc is expected to directly follow $ShipToLivorno$, and to directly precede $DeliverToCustomer$. Thus, the boolean expressions ShipToLivorno.M1 and not Exc.M1 and Exc.M1 and not DeliverToCustomer.M1 are associated to $Exc.PFG1$ and $DeliverToCustomer.PFG1$, respectively.

To detect when either $ShipToNaplesByRail$ or $ShipToNaplesBySea$ are executed, Exc is equipped with two Data Flow Guards identical to the ones of the two inner stages. To detect that the second leg was finished, Exc expects $ShipToNaplesByRail.M1$ to be achieved in December, or $ShipToNaplesBySea.M1$ to be achieved in the other months, and neither $ShipToNaplesByRail$ nor $ShipToNaplesBySea$ be active. Thus, the ECA rule if ((ShipToNaplesByRail.M1 and month[december]) or (ShipToNaplesBySea.M1 and not month[december])) and not (Active(ShipToNaplesByRail) or Active(ShipToNaplesBySea)) is defined for $Exc.M1$.

Finally, to ensure that the container should never overheat during the whole process, a new stage Seq, enclosing all the other stages, is introduced. A Fault Logger, which is triggered when the container becomes overheated, is then attached to Seq. To detect when the process starts, Seq is equipped with four Data Flow Guards identical to the ones of the inner stages. To detect that the process finished, Seq expects all the milestones of the inner stages to be achieved.

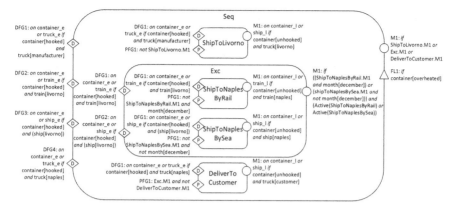

Fig. 4.5: E-GSM representation of the three-leg multi-modal shipment.

4.3 Assessing the severity of Constraints Violations

As mentioned in Section 4.2, the lifecycle of each stage is organized along three perspectives: status, outcome, and compliance. The status indicates if the activity associated to a stage is running. The outcome, on the other hand, determines if the activity associated to a stage was executed in the wrong way. Finally, the compliance determines if a stage satisfied the expected execution flow (i.e., the dependencies among the other stages).

This allows to determine, based on the current state of each stage, how severely the process is affected by discrepancies between the expected execution and the actual one. To do so, when the E-GSM process is modeled, a weight $S_i.w$, expressed as a positive integer, is assigned to each stage S_i composing the process: The more that weight is large, the more the associated stage is important for the process.

Then, when the process starts, a numeric value $S_i.sl$, named severity level, is assigned to each stage S_i. The criteria to assign $S_i.sl$ are described in Table 4.1, and are based on the state of the stages composing the process: If S_i was executed at the right time and its execution was successful, the severity level $S_i.sl$ is 0 (*none*). This is expected to be the normal behavior.

If S_i was incorrectly executed, but corrective action was taken (i.e., all the other stages that depended on S_i were not skipped), $S_i.sl$ is 1 (*low*). We consider 1 severer than 0 since an incorrect execution, although foreseen and properly handled, occurred.

If S_i was executed violating the dependencies among the other stages, or if the process is still running and S_i was skipped, $S_i.sl$ is 2 (*medium*). We consider 2 severer than 1 since a violation in the execution order of an activity is unforeseen and may unexpectedly impact the outcome of the process. On the other hand, as long as an incorrect execution is properly handled, the outcome of the process should not be affected.

Table 4.1: Severity levels that a stage S_i may get. $S_i.o$, $S_i.c$ and $S_i.s$ indicate the state of S_i along with the execution outcome, compliance and status respectively.

Severity level ($S_i.sl$)	Description	Condition
0	None	$S_i.o = regular \land S_i.c = onTime$
1	Low	$S_i.o = faulty \land \nexists S_j \mid (S_i \in S_j.DFG \land S_j.c = skipped)$
2	Medium	$S_i.c = outOfOrder \lor (S_i.c = skipped \land \exists S_j \mid S_j.s = opened)$
3	High	$S_i.c = skipped \land \nexists S_j \mid S_j.s = opened$
4	Critical	$S_i.o = faulty \land \exists S_j \mid (S_i \in S_j.DFG \land S_j.c = skipped)$

If the process terminated and S_i was skipped, $S_i.sl$ is 3 (*high*). We consider 3 severer than 2 since a skipped activity is more likely to affect the outcome of the process than an activity that was not executed at the right time. Also, we consider 3 severer than 1 since an unforeseen violation is more likely to affect the outcome of the process than a foreseen violation that was properly handled.

If S_i was incorrectly executed and no corrective action was taken (i.e., at least one of the stages that depended on S_i was skipped), $S_i.sl$ is 4 (*critical*). We consider 4 the severest issue, since it potentially combines the effects of both an unforeseen and a foreseen violation.

This way, it is possible to quantify the effects of a violation on each stage.

Given the severity level and the weight of each stage, it is then possible to assess the *health* of a process execution P_j, i.e., how much the deviations between the process model and the execution affect P_j:

$$Health(P_j) \to [0,1] = \frac{\sum_{S_i \in P_j} (S_i.w \cdot S_i.sl)}{\sum_{S_i \in P_j} (4 \cdot S_i.w)}$$

This way, if no constraint is violated, all stages composing the process have a severity level of 0, and the health of the process execution is 0 as well. On the other hand, if one or more constraints are violated, one or more stages have a severity level greater than 0. Consequently, the health of the process execution depends on how much the violation is severe (i.e. how big is the severity level of each affected stage), and how much are the affected stages important (i.e., how big is their weight).

It is then possible to define threshold values on the health of a process execution, in order to alert organization only when the process is critically affected by a deviation between the execution and the model.

4.4 E-GSM Expressiveness

Due to its declarative nature, E-GSM allows to introduce great flexibility in the processes being modeled with such a language. As previously discussed in Section 2.2.4, when deriving GSM models from imperative ones, control flow dependencies among activities are retained. Therefore, imperative constraints can also be modeled in GSM. Being an extension of GSM, E-GSM maintains this capability, which will be exploited in Chapter 5 to produce GSM models from BPMN collaboration diagrams.

In addition, E-GSM also supports declarative constraints. To give an idea of the potentials of E-GSM, this section presents how a set of declarative constructs, partially taken from [2], can be modeled in E-GSM. Due to time constraints, we were unable to provide an E-GSM counterpart for all the constructs described in [2]. Thus, we focused on those constructs that were the most relevant for our case studies in logistics. Although their behavior is quite simple to understand, these constructs are either complex or impossible to model with a traditional imperative language.

4.4.1 Activity Exclusion

As the default behavior of declarative languages is to allow every execution flow which is not explicitly specified in the model, those languages provide constructs to indicate actions that should not happen when the process is executed. The simplest one is probably the activity exclusion, meaning that the specified activity should never be executed when the process is run. It is worth noting that imperative languages do not offer such kind of constructs, since their default behavior is to forbid any execution flow which is not specified in the model. Therefore, by not indicating an activity in the model, an imperative language would forbid its execution throughout the whole process. However, in a monitoring context, if an activity is not explicitly specified in the model, it would be impossible for a monitoring platform to detect when such an activity is run, and consequently report that a violation occurred.

Fig. 4.6: Activity exclusion constraint.

Figure 4.6 shows how to model the activity exclusion constraint in E-GSM: It is sufficient to represent the activity with a stage S, and to attach to that stage a data flow guard $S.DFG_1$ and a milestone $S.M_1$ indicating under which conditions the activity is expected to start (i.e., $[S\_s]$) and stop (i.e., $[S\_t]$) its execution, respectively. Additionally, a process flow guard $S.PFG_1$ whose condition is always false is also attached to S.

This way, whenever the activity is executed, the corresponding stage is always be marked as *outOfOrder*.

4.4.2 Activity Overlap

Parallelism in imperative languages allows activities to be executed at the same time. However, it does not force any order constraint on the activities, meaning that they can be executed either in sequence or overlapped. E-GSM, on the other hand, allows to identify the execution as correct only if stages are always overlapped.

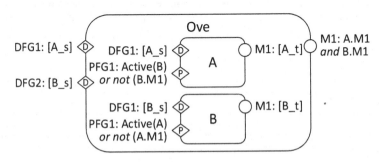

Fig. 4.7: Activity overlap constraint.

Figure 4.7 shows how to model the activity overlap constraint in E-GSM: The overlapped activities are represented as stages S_i enclosed in a stage *Ove*. *Ove* has a set *Ove.DFG* that includes all $S_i.DFG_k$, and a milestone *Ove.M$_1$* that requires that, for all S_i, at least one milestone $S_i.M_l$ be achieved. To each stage S_i, a process flow guard $S_i.PFG_1$ is added to check that at least one of the other stages S_j is *opened* ($\bigvee_{i \neq j}^{j} Active(S_j)$). Since this condition would be invalid for the stage that is firstly run (as no stage S_i would be *opened* at that moment), it is alternatively required that none of the milestones M_l of these stages be achieved ($\bigwedge_{i \neq j}^{j,l} S_j.M_l$).

This way, if an overlapped activity is executed after the other ones finished their execution (and their stages were marked as *closed*), its stage S_i is marked as *outOfOrder* (as the condition on $S_i.PFG_1$ was not true when S_i was opened). On the other hand, if an overlapped activity is not executed at all, *Ove* remains *opened* and, when the monitoring is stopped, the associated stage S_i becomes *skipped*.

4.4.3 Responded Existence

In imperative languages the execution order of activities is dictated by the control flow, which defines for each activity which are its direct predecessors and successors. However, in some cases it would be desirable to relax such execution order constraints, for example by imposing only that one activity B should follow another activity A, either directly or indirectly (i.e., if a third activity C directly follows A and B then follows C, the constraint would still be satisfied). This is easily achievable with declarative languages, which offer execution order constraints with various degrees of strictness. One of the less stringent execution order constraints is the responded existence: As stated by van der Aalst and Pesic in [2], if an activity A is being executed, then it must either be preceded or followed, either directly or indirectly, by another activity B. Note that the constraint holds even if B is executed and A is not, or if neither A nor B are executed.

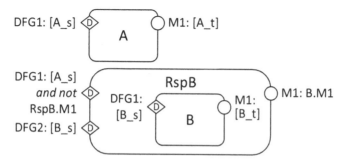

Fig. 4.8: Responded existence constraint.

Figure 4.8 shows how to model the responded existence constraint in E-GSM: The stage representing the dependent activity, which in this case is B, is enclosed in a stage $RspB$. $RspB$ has a milestone $RspB.M_1$ that requires at least one of the milestones of B be achieved, and a set of data flow guards $RspB.DFG$ that includes all $B.DFG_k$, plus the logic conjunction among the conditions of the data flow guards of the stage representing the triggering activity $A.DFG_k$, and $RspB.M_1$ not be achieved $((\bigcup^k B.DFG_k) \cup (\bigcup^k (RspB.M_1 \wedge c_k \in A.DFG_k)))$.

This way, if A is executed before B, $RspB$ is *opened* and, only when B completes its execution, it becomes *closed*. On the other hand, if B is executed before A, $RspB$ will not be *opened* again unless another execution of B takes place. By doing so, if $RspB$ is still *opened* when the monitoring is stopped, B is marked as skipped. It is worth noting that, once A is executed, the noncompliance on B can be assessed only when the monitoring is stopped, since B can potentially be executed throughout the whole execution of the process, thus fulfilling the responded existence constraint.

4.4.4 Constrained Iteration

While some imperative languages, and in particular BPMN, have constructs to indicate that an activity can be executed multiple times, they have either very stringent or no ways to limit the number of iterations for that activity. On the other hands, declarative languages allow to define an upper and/or a lower bound on the number of iterations.

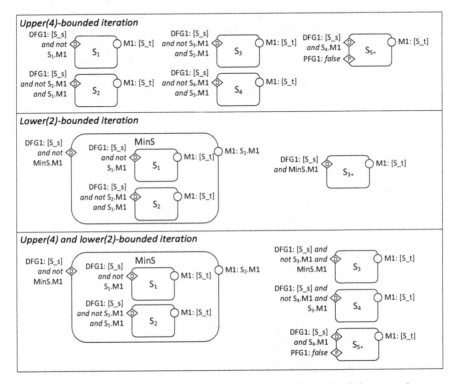

Fig. 4.9: Upper bounded (top), lower bounded (center), and both lower and upper bounded (bottom) iteration constraints.

The top portion of Figure 4.9 shows how to limit the number of iterations of an activity S to a maximum of m. S is represented by $m+1$ stages S_i. The milestones are identical for all these stages and predicate on the condition $[S\_t]$ indicating the termination of the activity ($\forall_{1 \leq i \leq m+1} S_i.M = [S\_t]$). The data flow guards, on the other hand, all predicate on the condition $[S\_s]$ indicating the activation of the activity but, in addition to that, also predicate on the milestones of the other activities: $S_1.DFG$ requires both $[S\_s]$ and the not achievement of any of $S_1.M$ ($[S\_s] \wedge \neg S_1.M$), $S_{m+1}.DFG$ requires both $[S\_s]$ and the achievement of any of $S_m.M$ ($[S\_s] \wedge S_m.M$), whereas $S_i.DFG$ where $1 < i < m+1$ requires

$[S\_s]$, the achievement of any of $S_{i-1}.M$, and the not achievement of any of $S_i.M$ ($[S\_s] \wedge S_{i-1}.M \wedge \not S_i.M$). Furthermore, a process flow guard whose condition is always false will be added to S_{m+1}.

This way, whenever the i-th iteration of S takes place, stage S_i is opened. If the number of iterations exceed m, stage S_{m+1} is opened and, since the condition on its process flow guard is always false, the stage is marked as *outOfOrder*.

The central portion of Figure 4.9, on the other hand, shows how to limit the number of iterations of S to a minimum of n. Even in this case, S is represented by $n+1$ stages S_i having all the same milestones, and $S_1.DFG$ and $S_i.DFG$ where $1 < i < n+1$ is derived as in the previous case. However, S_i where $1 \leq n < n+1$ is enclosed in a stage $MinS$, having data flow guards $MinS.DFG$ requiring both $[S\_s]$ and the not achievement of $MinS.M_1$, and a milestone $MinS.M_1$ requiring the achievement of any milestone $S_n.M$. Stage S_{n+1} has no process flow guard, and its data flow guards $S_{n+1}.DFG$ requires both $[S\_s]$ and the achievement of $MinS.M_1$.

This way, stage $MinS$ is *opened* when the first iteration of S takes place, but is closed only when the n-th iteration takes place. By doing so, if $MinS$ is still *opened* when the monitoring is stopped, all the stages representing the missing iterations of S will be marked as skipped. On the other hand, when the number of iterations is greater than n, their execution is captured by stage S_{n+1}, and $MinS$ will not be *opened* again. Even in this case, the noncompliance of S with respect to the minimum number of iterations can be assessed only when the monitoring is stopped, since further iterations of S could happen throughout the whole execution of the process.

Finally, by merging these two E-GSM representations, it is also possible to specify both a lower and an upper bound on the number of iterations that S should perform, as shown in the bottom portion of Figure 4.9.

4.5 Summary

This chapter presented E-GSM, an extension of the GSM artifact-centric language designed to autonomously and continuously monitor business processes. Stages, data flow guards and milestones keep track of which activities or process portions are running, were completed or are not yet started. Thank to process flow guards, it is possible to determine if the dependencies among stages (i.e., activities or process portions) are satisfied when the process is executed, opportunely flagging the stages that do not satisfy these dependencies as *outOfOrder* (executed when they should not) or as *skipped* (not executed when they should). Fault loggers allow to detect if something went wrong during the execution of a stage (i.e., an activity was incorrectly performed).

Thank to the declarative nature of E-GSM, relaxed dependencies among activities can be easily defined. To give an idea of this capability, it has been shown how some declarative constraints, that are normally impossible or difficult to model with an imperative language, can be easily modeled with E-GSM.

Finally, thank to the information that E-GSM collects when the process is run, whenever the execution deviates from the process model, it is possible to determine how severely each stage is affected by such a deviation. Based on the importance of each stage, it is then possible to assess how much deviations impact on the process, and determine the health of each execution.

Chapter 5
A Method to Easily Configure the Monitoring Platform

Due to the advantages we discussed in Section 4.2, we based our monitoring platform on E-GSM. Thank to E-GSM, it is possible to monitor the execution of a process along with two orthogonal perspectives: *(i)* the activities being executed and their dependencies, and *(ii)* the evolution of the artifacts. In fact, it may happen that, even though the process was correctly performed (i.e., all the activities were correctly executed at the right time), the artifacts were not manipulated as expected. On the other hand, even though the artifacts were correctly manipulated, a process may be incorrectly executed (e.g., an activity was not executed even though it was mandatory). Therefore, to address i, an E-GSM model representing the activities composing the process and their dependencies is required. Another E-GSM model, representing all the admissible states and transitions for an artifact, is required to address ii.

As discussed in Section 3.3, the smart objects impersonating the artifacts, which coincide to the physical objects participating in a process, are responsible for monitoring the process. Consequently, when more than one artifact participate in the process, the smart objects must exchange information on their state with each other. Therefore, smart objects must know the identity of the other ones participating in the same process execution. As the identity may be known only after the process start, the binding between the smart objects objects and the process execution must be definable at runtime. Also, at some point in time smart objects may no longer participate in the same process execution. As such, information on their state should no longer be notified to the other smart objects. Therefore, mechanisms to unbind smart object and the process execution should be supported as well. To support such binding and unbinding mechanisms, our platform relies on the Events Router component. This component, in turn, relies on the events notified by the organizations participating in the process. Criteria that associate an event to the binding or unbinding of a smart object, named mapping criteria, have to be provided to the Events Router before the monitoring starts.

To help process designers in configuring our monitoring platform, we propose a method that, starting from the well-known BPMN 2.0 collaboration diagrams [29], derives the E-GSM models and the mapping criteria in a semi-automatic way. This

© Springer Nature Switzerland AG 2019
G. Meroni: Artifact-Driven Business Process Monitoring, LNBIP 368, pp. 61–92, 2019
https://doi.org/10.1007/978-3-030-32412-4_5

way, process designers do not have to learn E-GSM to monitor a process, and can reuse existing process models and modeling tools.

5.1 Steps

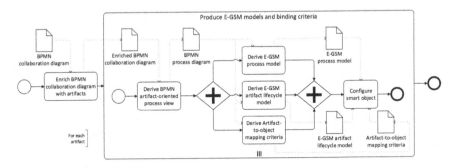

Fig. 5.1: Overview of the transformation process.

Figure 5.1 shows the main steps characterizing our method. The starting point is a multi-party process modeled with a BPMN collaboration diagram, such as the one shown in Section 3.1, that specifies the behavior of the involved organizations and the communication among them in terms of exchanged messages.

The first step consists in enriching this diagram with information on the artifacts participating in the process and their state. This way, it is possible to indicate which artifacts are required for an activity to start, and how such an activity alters the artifacts. Then, for each artifact, the E-GSM models and the mapping criteria to drive the monitoring platform running on the smart object embodying such an artifact are derived.

To do so, *(i)* the artifact-oriented view of the process (i.e., the portion of the process relevant for that artifact) is extracted, then *(ii)* the E-GSM process model, representing the activities composing the process to monitor and the dependencies among activities, is generated, *(iii)* the E-GSM artifact lifecycle model, capturing the lifecycle of the artifact, is created, and *(iv)* the artifact-to-object mapping criteria, specifying when other artifacts start and stop interacting with the process, are produced.

Once the E-GSM models and the mapping criteria are produced, the smart object impersonating the artifact can be configured to monitor the process.

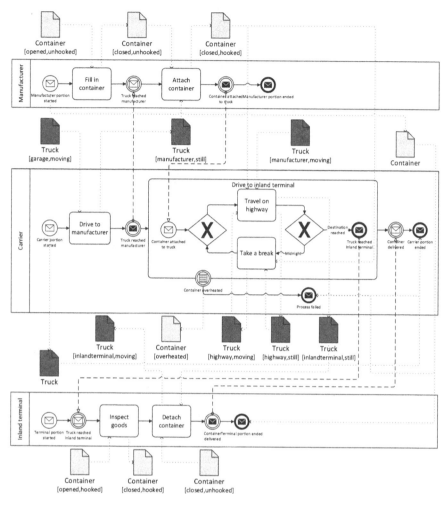

Fig. 5.2: BPMN collaboration diagram enriched with information on the artifacts that participate in the process.

5.1.1 Enriching the BPMN Process Model With Artifacts

A BPMN collaboration diagram represents the activities carried out by the organizations participating in the process, which are represented with the *pool* construct, and their execution flow dependencies, represented with the *gateway, sequence flow*, and *message flow* constructs. However, to rely on smart objects for detecting when activities are executed, the diagram must also capture *(i)* which artifacts are required for an activity to be executed, *(ii)* which state should the artifacts assume for an activity to start, and *(iii)* how the execution of an activity affects the state of the artifacts. Additionally, to automatically determine when a smart object should be bound or unbound to the process, the diagram should indicate *(iv)* when an artifact starts interacting with the process, *(v)* how the identity of the smart object impersonating the artifact is notified to the process, and *(vi)* when an artifact is no longer related to the process.

To model this information, we rely on the standard BPMN *data objects*, rather than introducing yet another extension of BPMN. Data objects traditionally serve for documentation purposes, yet we use them to model the artifacts and their interactions with the process. This way, process designer can perform this step by using any existing modeling tool, as long as it conforms to the BPMN 2.0 standard[1]. Moreover, we establish the following set of rules to guarantee at design time that the BPMN model contains enough information to completely and unambiguously automate the monitoring of the process at runtime.

- An artifact must be modeled with data objects. The name of the data object identifies the artifact (e.g., Truck), whereas the data state identifies in which condition the artifact is supposed to be (e.g., [highway,moving]).
- Each monitored activity must have at least one input data object with a specified data state. The activity is supposed to start only when all input data objects exist and have the specified data state. If an activity has two input data objects referring to the same artifact in different data states, the artifact must assume one of the specified states.
  For example, Travel on highway starts when Truck is either in state [highway,moving], or in [manufacturer,moving], and Container is in state [closed,hooked].
- Each activity must have at least one output data object with a specified data state. The activity is supposed to finish once all output data objects exist and have the specified data states. If an activity has two output data objects that refer to the same artifact in different data states, the artifact must assume one of the specified states.
  For example, Travel on highway ends when Truck is either in state [highway,still] or in [inlandterminal,still].
- For each artifact, at least one output data object with no data state must be defined in the diagram and associated to a generic, message or signal start or intermediate catch event. The artifact is supposed to begin interacting with the process when that event occurs. Beforehand, the artifact and its state is ignored. The payload of the event indicates the identity of the smart object that instantiates the artifact.
  In the example, Container starts interacting when the process portion carried out by the manufacturer starts. Truck is bound when the process portion carried out by the carrier starts.
- For each artifact, zero or more input data objects with no data state can be defined in the diagram and associated to a generic, message or signal end or intermediate throw event. The artifact is supposed to become unrelated to the

---

[1] For the running example, we used Eclipse BPMN2 Modeler `http://www.eclipse.org/bpmn2-modeler`, Camunda Modeler `https://camunda.org/download/modeler`, and Bizagi BPM Modeler `https://www.bizagi.com/en/products/bpm-suite/modeler`.

process when the event occurs. After such an event, the artifact and its state will be ignored when the process is executed.

For instance, Truck will be no longer related to the process once the process portion carried out by the carrier ends.

- Data associations must not contradict the semantics of the sequence flow as they are used to identify when activities start or end. Therefore, care should be taken when associating artifacts to the activities. In particular, each combination of artifacts and their states should not be associated to multiple activities, unless they are expected to be executed simultaneously. Similarly, combinations should be chosen to ensure that they should occur only when the associated activity should start, and not in other occasions.

  For example, Travel on highway cannot be declared to start only when Truck is in [manufacturer,moving], otherwise it could not start again after a break along the journey, far from the inland terminal, is taken (despite the loop in the process model). Therefore Truck[highway,moving] is set as another input for Travel on highway and as an output of Take a break.

- If two activities are carried out in parallel, output data objects that refer to the same artifact cannot be associated to both activities. Otherwise, the lifecycle of the artifact would be non-deterministic.

Example.

Figure 5.2 shows the process model obtained by extending the one presented in Section 3.1 according to the previously mentioned rules. The input and output data objects of Detach container indicate the preconditions and postconditions for that activity to be executed. To execute Detach container, the container must be closed, hooked to the truck and not overheated, and the truck must already be parked on the inland terminal's premises. When Detach container finishes, the container will be unhooked from the truck. As the container is not required when the truck drives to the manufacturer's premises, its identity can be indicated before, during, or after that activity is performed. On the other hand, the identity of the container is required for activity Fill in container to be performed. Hence, the Container data object is associated to the start event of the manufacturer's process portion.

5.1.2 Extracting the Artifact-oriented Process View

A BPMN collaboration diagram, with the pool and lane constructs, allows to define which organization and business unit is in charge of executing each activity. Additionally, with the sequence and message flow constructs, it makes possible to distinguish, respectively, intra-organizational dependencies among activities from inter-organizational ones. However, as an artifact can travel along different organizations and participate in activities carried out by different organizations, it no longer makes sense to distinguish activities and dependencies based on the organizations. Additionally, from an artifact's point of view, activities and events that neither require nor alter the artifact are irrelevant for its evolution. Therefore, they should be excluded.

Thus, to model a process from the view point of an artifact, *(i)* activities and events that do not interact with the artifact should be excluded, *(ii)* activities should no longer be confined inside single organizations, and *(iii)* no distinction among message flow and sequence flow should be made. Instead of defining this new process model from scratch, designers can derive it from the BPMN collaboration diagram resulting from the previous step. To do so, we defined a set of steps that, given an artifact Ar, transform the enriched collaboration diagram into a process diagram representing the process from the viewpoint of the artifact:

- To fulfill i, elements not interacting with Ar are removed:

    – Activities that have at least an input or an output data object that refers to Ar are kept, while other activities are removed. This way, only those activities that interact with Ar are considered, while the other ones are excluded.
    – Events that have one input or one output data object that refers to Ar are kept, as long as the state of Ar is specified (i.e., the event is not responsible for indicating when Ar starts or stops interacting with the process). This way, events that influence or are influenced by Ar are considered.
    On the other hand, events that are responsible for notifying when the artifact starts or stops interacting with the process are excluded. The rationale behind this choice is that, being the monitoring performed directly by smart objects impersonating the artifacts, each smart object already knows its own identity, and is active only when the process is performed. Consequently, these events are irrelevant for the monitoring, and they can be ignored.
    – Events that have one input or one output data object that refers to an artifact Ar' (different from Ar) without state are kept, as long as at least one of the input or output data objects of the kept activities refers to Ar' as well (i.e., the event is responsible for indicating when Ar' starts or stops interacting with the process). This way, events that influence or are influenced by artifacts different than Ar are excluded.
    On the other hand, events that are responsible for notifying the identity of the other artifacts are kept. This is necessary because, when an activity requires both Ar and another artifact Ar' to assume a specific state to be executed, the identity of those artifacts is needed. Since monitoring is directly performed on smart objects impersonating the artifacts, the smart object impersonating Ar already knows its own identity, whereas the identity of Ar' has to be notified. Also, the smart object needs to know when Ar' is no longer related to the process. Therefore, events responsible for binding and unbinding Ar' to the process are retained.
    – Events that have incoming or outgoing message flows are kept. This way, the coordination between pools is preserved.
    – All other events are removed.

– Empty subprocesses (i.e., subprocesses whose internal activities and events were entirely removed) are removed.

- To fulfill ii, pools are removed. This way, the model becomes a BPMN process diagram.
- To fulfill iii, message flows are replaced with sequence flows.

  – Message throw events with an outgoing message flow (i.e., responsible for coordinating pools) are replaced with parallel split gateways. The rationale behind this choice
  – Message catch events with an incoming message flow (i.e., responsible for coordinating pools) are replaced with parallel merge gateways.
  – If the message flow ends to a start event inside a subprocess activity, the sequence flow ends to the subprocess activity.
  – If the message flow starts from an end event inside a subprocess activity, the sequence flow starts from the subprocess activity.
  – If the message flow starts from or ends to an intermediate event inside a subprocess activity, the subprocess is removed.

- To ensure that the diagram is well-formed and, except for the absence of activities and events not related to Ar, dependencies among activities are as closest as possible to the ones in the original BPMN collaboration diagram, the following modifications are performed:

  – When an activity or an intermediate event is removed, a sequence flow connecting its predecessor to its successor is introduced.
  – Boundary blocking events that were attached to a removed activity are replaced with exclusive split gateways with no branch condition.
  – Boundary non-blocking events that were attached to a removed activity are replaced with inclusive split gateways with no branch condition.
  – Data objects that were connected to a removed activity or event are removed, as long as they are not connected to any other activity or event that is kept.
  – If the resulting model has no start event, a generic start event is added and connected to the elements that have only outgoing sequence flows.
  – If the resulting model has no end event, a generic end event is added and connected to the elements that have only incoming sequence flows.

These steps, as shown in Figure 5.3, can be formalized with an algorithm. This way, once the designer completes enriching the BPMN collaboration diagram with information on the artifacts, he can automatically obtain a process view for each artifact in the model. However, these automatically generated process diagrams may not be well-structured, which is a prerequisite for the following steps. According to Reichert and Weber in [112], a process model is well-structured if it is made of process-portions, named blocks, that have a single inbound sequence flow and a single outbound sequence flow, that can be nested, and that must not overlap. Therefore, the designer still has to manually inspect the automatically generated diagrams, and eventually modify them to make them well-structured.

```
1: function PRODUCEPROCESSVIEW(collaborationDiagram,artifact)
2:     procView ← copy(collaborationDiagram);              ▷ duplicate the source BPMN model
3:     for all i = 1: procView.elements.count do ▷ examine each element in the BPMN model
4:         keep ← false;
5:         if            (procView.elements.get(i).type              ='Activity')        ∨
     (procView.elements.get(i).type ='Event') then
6:             if (artifact ∈ procView.elements.get(i).inputDataObjects) ∨ (artifact ∈
     procView.elements.get(i).outputDataObjects) then        ▷ element references the artifact
7:                 keep ← true;
8:             end if
9:             if            (procView.elements.get(i).type            ='StartEvent')       ∨
     (procView.elements.get(i).type ='EndEvent')∨(procView.elements.get(i).type ='IntermediateEvent')
     then
10:                if          (procView.elements.get(i).msgTo            ≠null)        ∨
     (procView.elements.get(i).msgFrom ≠null) then        ▷ element is not responsible for the
     collaboration
11:                    keep ← true;
12:                end if
13:            end if
14:         end if
15:         if keep = false then                          ▷ make discarded element an orphan
16:             procView.elements.get(i).predecessor.successor                          ←
     procView.elements.get(i).successor;
17:             procView.elements.get(i).successor.predecessor                          ←
     procView.elements.get(i).predecessor;
18:             procView.elements.get(i).predecessor ← null;
19:             procView.elements.get(i).successor ← null;
20:         end if
21:     end for
22:     removePools(procView);                    ▷ remove pools from process model
23:     removeOrphans(procView);               ▷ remove discarded elements (i.e., orphans)
24:     makeWellformed(procView);        ▷ make target model well-formed (if not already so)
25:     return procView;
26: end function
```

Fig. 5.3: (Simplified) algorithm to produce, given a BPMN collaboration diagram and an artifact, the process view

Example.

Figure 5.4 shows the BPMN process diagrams obtained from the BPMN collaboration diagram shown in Figure 5.2 by following the guidelines. The diagram shown in the top portion of Figure 5.4 represents the process from the container's viewpoint, whereas the one shown in the bottom portion of Figure 5.4 represents the process from the truck's viewpoint.

For the container, activity Drive to manufacturer is removed since it does not interact with the container. Events Terminal portion started and Manufacturer portion ended are removed, as they have no data object associated. Data object Truck[garage,moving] is removed, since none of the remaining activities uses it. Data object Container, together with events Manufacturer portion started and Terminal portion ended, are also removed, as the container that will perform the monitoring

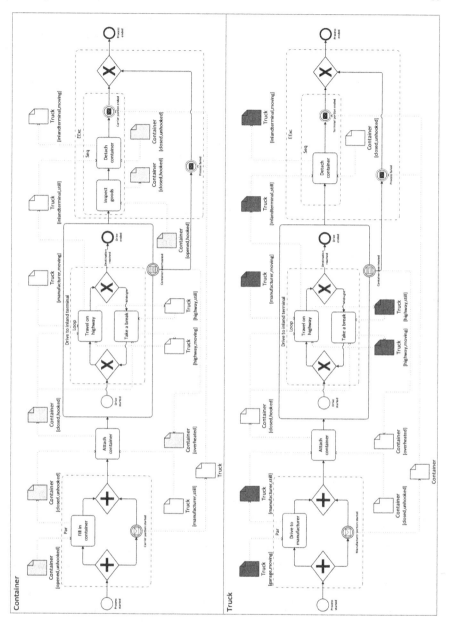

Fig. 5.4: BPMN process diagram representing the process from the viewpoint of the container (top) and the truck (bottom).

already knows its own identity. On the other hand, data object Truck, together with events Carrier portion started, Carrier portion ended, and Process failed, are kept, as the container needs to know the identity of the truck to monitor activities that are also linked to that artifact.

Likewise, for the truck, activities Fill in container and Inspect goods, data objects Container[opened,unhooked], Container[opened,hooked] and Truck, and events Carrier portion started, Carrier portion ended, Terminal portion started, and Manufacturer portion ended are removed.

5.1.3 Generating the E-GSM Process Model

To instruct the monitoring platform to determine if the process is executed according to the model, an E-GSM model that captures the activities composing the process and their dependencies has to be derived from the BPMN process diagram obtained in the previous step. To lift designers from learning E-GSM and manually producing the E-GSM model, we defined a set of rules to automatically transform BPMN constructs and patterns in their E-GSM counterparts. These rules are extensively discussed in [84], and have been formalized in ATLAS Transformation Language (ATL) [58]. This way, such a transformation can be fully automated, as long as the source BPMN process diagram satisfies the following conditions:

- **DR1**: The process diagram is well-structured. Since the transformation rules rely on block nesting to preserve control flow dependencies, this condition is required for the transformation to work.
- **DR2**: The process diagram does not contain multiple-instance activities. As stated in [35], "each [GSM] stage can have at most one active occurrence at a given time". Currently, E-GSM retains this limitation. Thus, multiple-instance activities must be rendered as process portions. Otherwise, they are treated as basic tasks.
- **DR3**: The process diagram does not contain link events, as these events do not really occur when the process is run. Thus, the designer has to remove them by manually merging process diagrams.

If the source diagram satisfies these conditions, an E-GSM model suited for our monitoring platform can be automatically produced. It is worth noting that, being the resulting E-GSM model automatically derived from a BPMN one, it cannot fully exploit the expressiveness of the E-GSM formalism. Its expressiveness is in fact limited by the one of the BPMN formalism, except for the sequence flow dependencies that are no longer prescriptive and can be violated when the process is executed. As such, purely declarative constraints, such as the ones presented in Section 4.4 cannot be present in the resulting E-GSM model, as they cannot be modeled in BPMN. However, in case such a flexibility is required for the monitoring, the designer can alter the E-GSM model and manually introduce such constraints, a task which is still easier than redesigning the process in E-GSM from scratch.

To make this book self-contained, in the remainder of this section we present a subset of these transformation rules.

5.1.3.1 Basic Elements

The transformation rules defined for basic elements are presented in Figure 5.5.

Rule A BPMN *task* activity A is translated into a *stage* S with one *data flow guard* $S.DFG_1$ and one *milestone* $S.M_1$.
The ECA rule that defines $S.DFG_1$ depends on the input data objects of A:

- If A has at least one input data object, the ECA rule is triggered when a change in the state of one of the artifacts Ar associated with each input data object of A occurs, and Ar enters the current state (which generates event Ar_e). It will only be fired if the state assumed by all artifacts Ar is the one indicated by the input data objects of A.
- If A has no input data object, the activation of A has to be manually notified. Therefore, the ECA rule is fired by the occurrence of event A_s.

The ECA rule that defines $S.M_1$ depends on the output data objects of A:

- If A has at least one output data object, the ECA rule is triggered when a change in the state of one of the artifacts Ar associated with each output data object of A occurs, and Ar leaves the previous state (which generates event Ar_l). It will only be fired if the state assumed by all artifacts Ar is the one indicated by the output data objects of A.
- If A has no output data object, the termination of A has to be manually notified. Therefore, the ECA rule is fired by the occurrence of event A_t.

Rule A BPMN *start, end* or *intermediate event* e is translated into a *stage* S' with one *data flow guard* $S'.DFG_1$ and one *milestone* $S'.M_1$.
The ECA rule that defines $S'.DFG_1$ and $S'.M_1$ depends on the nature of the event:

- For *generic, message* and· *signal events*, the ECA rule is fired by the occurrence of the event.
- For *condition events*, the ECA rule is fired when a change in the state of the artifact Ar associated to the output data object of e occurs (which generates event Ar_l), and the state assumed by Ar is the one indicated by the data object of e.
- For *timer events*, the ECA rule is fired when the constraint on the time is fulfilled.

Rule A BPMN *activity* A with a *non-interrupting boundary event* e attached is translated into a *stage* S according to Rule 1 with one *fault logger* $S.FL_1$.
The ECA rule that defines $S.FL_1$ depends on the nature of the event:

- For *generic, message* and *signal events*, the ECA rule is fired by the occurrence of the event.
- For *condition events*, the ECA rule is fired when a change in the state of the artifact Ar associated to the output data object of e occurs (which generates event Ar_l), and the state assumed by Ar is the one indicated by the data object of e.

- For *timer events*, the ECA rule is fired when the constraint on the time is fulfilled.

Rule A BPMN *activity A* with an *interrupting boundary event e* attached is translated into a *stage S* according to Rule 1 with one additional *milestone S.Me* and one *fault logger S.FL$_1$*.

The ECA rule that defines *S.Me* and *S.FL$_1$* depends on the nature of the event:

- For *generic, message* and *signal events*, the ECA rule is fired by the occurrence of the event.
- For *condition events*, the ECA rule is fired when a change in the state of the artifact *Ar* associated to the output data object of *e* occurs (which generates event *Ar$_l$*), and the state assumed by *Ar* is the one indicated by the data object of *e*.
- For *timer events*, the ECA rule is fired when the constraint on the time is fulfilled.

Figure 5.6 shows an excerpt of the ATL code implementing these rules[2]. In particular, rule `activity2substage` is responsible for transforming each activity in the source BPMN process diagram into a *stage* in the target E-GSM process model. `activity2substage` also produces, for each activity, a *process flow guard* that is attached to the corresponding *stage* (the boolean expression associated to the *process flow guard* is determined by invoking the helper function `producePFGExpression`).

Lazy rules `produceDFG`, `produceM` and `produceFL` are also defined to dynamically generate, respectively, *data flow guards*, *milestones*, and *fault loggers*. When `activity2substage` is triggered, `produceDFG` and `produceM` are invoked once, and their ECA is determined by the input and output data objects associated to the activity. Then, `produceM` is invoked as many times as the number of interrupting boundary events associated to the activity, and `produceFL` as the number of boundary events (both interrupting and non-interrupting), and their ECA predicates on the occurrence of the boundary event.

5.1.3.2 Normal Flow

To maintain execution flow dependencies among activities and events in the target E-GSM model, the source BPMN process diagram is decomposed into nested blocks. In particular, starting from the classical control flow patterns [117], five types of blocks are identified according to their structure:

- A *sequence block* is made of linked activities, events and other blocks without splits or merges. It corresponds to pattern *sequence*.
- A *parallel block* organizes activities, events, and other blocks in two or more parallel threads resulting from the combination of patterns *parallel split* and *synchronization*.

---

[2] For the sake of clarity, in this paper we report only a simplified excerpt of the source code. A complete version is publicly available at `https://bitbucket.org/polimiisgroup/bpmn2egsm/src/206dd0270c4f32a7997d847356b0397fb283aac4/BPMN2GSM/BPMN2XGSM.atl`.

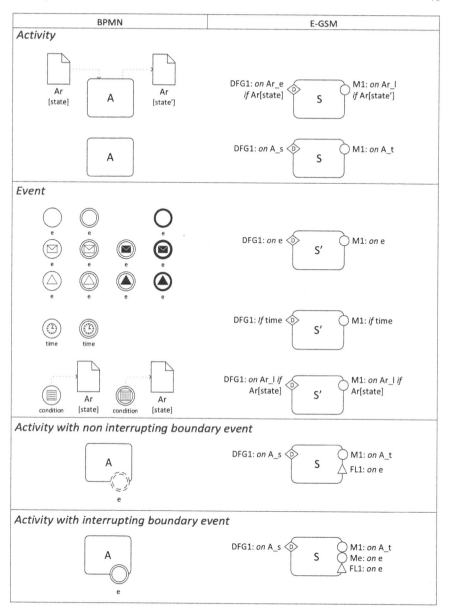

Fig. 5.5: BPMN to E-GSM transformation rules for basic elements.

```
1    lazy rule produceDFG {
2     from expr: String, stage: String
3     to rel: XGSM!DataFlowGuardType (
4       id <- stage+'_dfg1',
5       expression <- expr )}
6
7    lazy rule produceM {
8     from expr: String, stage: String
9     to res: XGSM!MilestoneType (
10      id <- stage+'_m1',
11      expression <- expr )}
12
13   lazy rule produceFL {
14    from expr: String, stage: String
15    to rel: XGSM!FaultLoggerType (
16      id <- stage+'_fl1',
17      expression <- expr )}
18
19   rule activity2substage {
20    from s: BPMN!Activity
21    to
22     tpfg: EGSM!ProcessFlowGuardType (
23       id <- s.id+'_pfg',
24       expression <- s.producePFGExpression() ),
25     tss: EGSM!SubStageType (
26       id <- s.id,
27       processFlowGuard <- tpfg
28       dataFlowGuard <- OrderedSet{produceDFG(s.id, 'on '+'or'.concatStrings(
29        self.dataInputAssociations -> iterate (g; ret: OrderedSet(String)
30        = OrderedSet{} | ret -> including(g.sourceRef.name+'_e'))+' if '+ret
31        -> including('('+'or'.concatStrings(self.dataInputAssociations
32        .-> iterate(f; ret: OrderedSet(String) = OrderedSet{} |
33         if (f.sourceRef.name=e.sourceRef.name) then
34          ret -> including(f.sourceRef.name+'['+
35           f.sourceRef.dataState.name+']')
36         else
37          ret
38         endif
39        ))+')')))},
40       milestone <- OrderedSet{produceM(s.id, 'on '+'or'.concatStrings(
41        self.dataOutputAssociations -> iterate (g; ret: OrderedSet(String) =
42        OrderedSet{} | ret -> including(g.targetRef.name+'_1'))+' if '+ret
43        -> including('('+'or'.concatStrings(self.dataOutputAssociations
44        -> iterate(f; ret: OrderedSet(String) = OrderedSet{} |
45         if (f.targetRef.name=e.targetRef.name) then
46          ret -> including(f.targetRef.name+'['+
47           f.targetRef.dataState.name+']')
48         else
49          ret
50         endif
51        ))+')' ))} -> including(s.getBlockingBoundaryEvents()
52        -> collect(b | produceM(s.id+'_'+b.id, 'on '+b.id)))),
53      faultLogger <- s.boundaryEventRefs ->
54       collect(e | produceFL(s.id+'_'+e.id ,'on '+e.id))
55     )
56   }
```

Fig. 5.6: (Simplified) ATL translation rule to derive from a BPMN Activity the corresponding E-GSM construct

- A *conditional exclusive block* organizes activities, events, and other blocks in two or more branches resulting from a combination of patterns *exclusive choice* and *simple merge*.
- A *conditional inclusive block* organizes activities, events, and other blocks in two or more branches resulting from a combination of patterns *multi-choice* and *structured synchronized merge*.
- A *loop block* organizes activities, events, and other blocks according to pattern *structured loop*.

For each of these blocks, a transformation rule is defined to produce E-GSM constructs that maintain the same behavior. A graphical representation of these rules is reported in Figure 5.7.

Rule A *sequence block* corresponds to a *stage Seq* that includes S_x inner *stages* obtained by applying the transformation rules to all the elements (i.e., *activities*, *events*, inner *blocks*) that belong to the block.

- In addition to the existing *process flow guards*, each inner *stage* has $S_x.PFG1$ to state that none of its *milestones* is achieved, and at least one of the *milestones* of the element that directly precedes it (if present) is achieved. This way, inner *stages* are expected to be opened only once, and only after their direct predecessor is closed.
- *Seq* has a *set of data flow guards Seq.DFG* that includes all $S_x.DFG_i$, and a *milestone Seq.M1* that requires that, for all inner *stages* S_x, at least one *milestone* $S_x.M_j$ be achieved. This way, *Seq* is opened when at least one of its inner *stages* S_x is opened too, and – as achieving a *milestone* is enough to close a *stage* – *Seq* is closed when all inner *stages* S_x are closed. □

It is worth noting that, being the process well-structured, we can always identify a single *sequence block* that covers the entire process definition. This *sequence block* corresponds to the sequence flow that initiates with the start event, ends with the end event, and traverses any other activity, intermediate event, or internal block, if present. This way, each BPMN *subprocess activity* can be mapped to the *stage* derived from the *sequence block* which covers that specific subprocess completely.

Rule A *parallel block* corresponds to a *stage Par* that includes all the *stages* obtained by applying Rule 5 to all its threads, which result in S_x inner *stages*.

- In addition to the existing *process flow guards*, each inner *stage* has $S_x.PFG1$ to state that none of its *milestones* is achieved. This way, inner *stages* are expected to be opened only once.
- *Par* has a *set of data flow guards Par.DFG* that includes all $S_x.DFG_i$, and a *milestone Par.M1* that requires that, for all inner *stages* S_x, at least one *milestone* $S_x.M_j$ be achieved. □

Rule A *conditional exclusive block* corresponds to a *stage Exc* that includes all the *stages* obtained by applying Rule 5 to all its branches, which result in S_x inner *stages*.

- For each *stage* S_x, a *process flow guard* $S_x.PFG1$ is added to check that no *milestone* $S_x.M_j$ has already been achieved, that the condition on the branch from which S_x is produced (if present) is satisfied, and that none of the other inner *stages* is opened (i.e., not Active(S_y) where y≠x). This way, S_x are expected to be opened only once, and only when their branch is taken and no other branch is. Only one branch is expected to be active.
- *Exc* has a *set of data flow guards Exc.DFG* that includes all $S_x.DFG_i$, and a *milestone Exc.M1* that requires that, for at least one *stage* S_x, one *milestone* $S_x.M_j$ be achieved, and the condition on the branch from which S_x is produced (if present) be satisfied, as long as none of the other inner *stages* is opened. This way, *Exc* is opened when at least one of its inner *stages* can be opened too, and closed when the activated inner *stages* become closed, as long as no other *stage* is opened. □

Rule A *conditional inclusive block* corresponds to a *stage Inc* that includes all the *stages* obtained by applying Rule 5 to all its branches, which result in S_x inner *stages*.

- For each *stage* S_x, a *process flow guard* $S_x.PFG1$ is added to check that no *milestone* $S_x.M_j$ has already been achieved, and that the condition on the branch from which S_x is produced (if present) is satisfied. This way, S_x are expected to be opened only once, and only when their branch is taken. Multiple branches can be active at the same time.
- *Inc* has a *set of data flow guards Inc.DFG* that includes all $S_x.DFG_i$, and a *milestone Inc.M1* that requires that, for at least one *stage* S_x, one *milestone* $S_x.M_j$ be achieved, and the condition on the branch from which S_x is produced (if present) be satisfied, as long as none of the other inner *stages* is opened. □

Rule A *loop block* corresponds to two *stages*, *Ite* and *Loop*. *Ite* includes all the inner *stages* S_x that are obtained by applying Rule 5 to all the branches within the *loop block*. One of these *stages* is a *forward stage*, that is, its sequence flow goes in the same direction as the sequence flow that includes the loop block. The others are *backward stages*.

- For all the inner *stages*, a *process flow guard* $S_x.PFG1$ is added to check that no *milestone* $S_x.M_j$ is already achieved. Moreover, if S_x is a *backward stage*, $S_x.PFG1$ also requires that the condition on the branch (if present) be satisfied, and that one of the *milestones* of the *forward stage* be achieved. This way, both *stages* are expected to be opened only once and, for the *backward stages*, only after the *forward stage* is closed, the branch they represent is taken, and no other branch is.
- *Stage Ite* has a *set of data flow guards Ite.DFG* that includes all $S_x.DFG_i$, and two *milestones*, *Ite.M1* and *Ite.M2*, where:

  - *Ite.M1* requires that one of the *milestones* of the *forward stage* be achieved and the exit condition of the loop (if present) be satisfied, as long as no *backward stage* is opened.

– *Ite.M2* requires that one of the *milestones* of the *forward stage* be achieved and, for at least a *backward stage*, one of its *milestones* be achieved and the condition on that branch (if present) be satisfied, as long as none of the other *backward stages* is opened. □

Ite has no *process flow guards* since it is supposed to be executed multiple times and, every time it becomes opened, a new iteration of the loop is carried out. Thus, *Ite* is opened when at least one of its inner *stages* can be opened too, and it is closed when either the process can exit the loop (i.e., *Ite.M1* is achieved), or when an iteration is complete (i.e.,*Ite.M2* is achieved).

• *Stage Loop* includes *Ite* and has a *set of data flow guards Loop.DFG = Ite.DFG*, and a milestone *Loop.M* that requires that milestone *Ite.M1* be achieved. This way, the process can exit the loop when the exit condition (if present) is satisfied.

5.1.3.3 Exceptional Flow

BPMN supports the management of foreseen exceptions through *boundary events*, that is, events directly attached to activities. These events, like split gateways, determine a branching of the sequence flow into an *exceptional flow*, which leaves the boundary event, and a *normal flow*, to continue the execution from the activity. If the foreseen exception occurs while executing the activity, the attached boundary event activates the exceptional flow. A dedicated set of rules shown in Figure 5.8 is thus required to preserve this behavior in the resulting E-GSM model.

Interrupting boundary events cause the normal and exceptional flows to be mutually exclusive, therefore we expect them to be merged by a BPMN *exclusive merge gateway* at the end. This requires that two additional blocks, called *forward exception handling* and *backward exception handling*, respectively, be defined. The *forward exception handling block* comprises an *interrupting boundary event*, and a *simple merge* control flow pattern, defined with a BPMN *exclusive gateway*, that merges the exceptional flow and the portion of the normal flow that *follows* the activity to which the boundary event is attached. Its behavior is similar to the one of the *conditional exclusive block*, with the exception of the branch condition, which predicates on the achievement of the milestone derived from the boundary event.

The *backward exception handling block* also comprises an *interrupting boundary event* and a simple merge control flow pattern. However, the BPMN exclusive gateway composing that pattern merges the exceptional flow and the portion of the normal flow that *precedes* the activity to which the boundary event is attached. This block produces a loop that allows one to re-execute part of the normal flow if the boundary event is triggered, and therefore it is translated similarly to a *loop block*.

In BPMN, boundary events could also be non interrupting, that is, they activate the exceptional flow without terminating the associated activity. Therefore, the elements within the exceptional flow can run in parallel with the normal flow that starts from the activity the boundary event is associated with. Since we expect

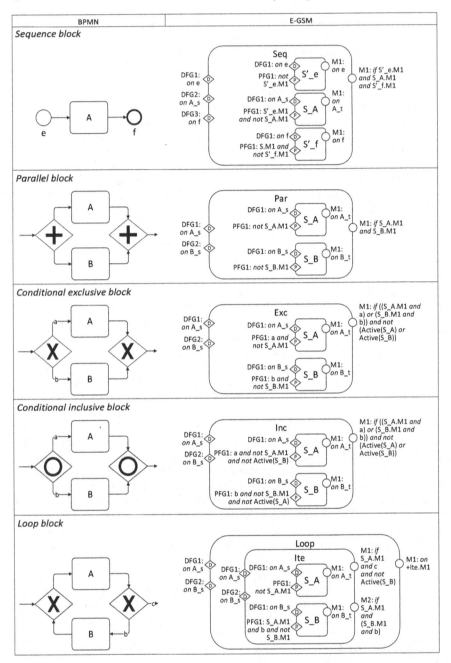

Fig. 5.7: BPMN to E-GSM transformation rules for normal flow blocks.

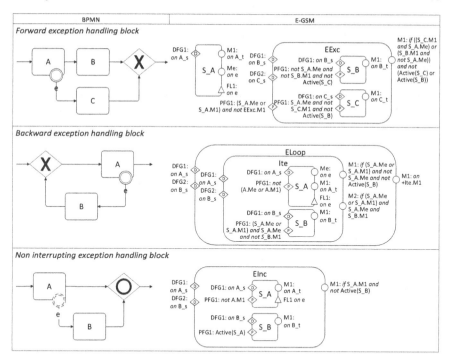

Fig. 5.8: BPMN to E-GSM transformation rules for handling exceptions.

these potentially simultaneous sequence flows be merged by an inclusive merge gateway, the transformation requires an additional block, called *non interrupting exception handling block*. This new block comprises a *non interrupting boundary event* to split the execution flow into an exceptional flow and the continuation of the normal one, and a *structured synchronized merge*, defined with a BPMN *inclusive gateway*, to merge the two flows in case the exception occurred.

Example.

Figure 5.9 shows the E-GSM process model derived from the BPMN diagram shown in the top portion of Figure 5.4, representing the process from the container's viewpoint. *AttachContainer.DFG1* is triggered whenever artifacts Truck or Container enter a new state (which is represented by $truck_e$ or $container_e$). Furthermore, *AttachContainer.DFG1* requires that Truck and Container are in states [manufacturer,still] and [closed,unhooked], respectively. *AttachContainer.M1*, on the other hand, is triggered whenever the container leaves the current state ($container_l$), and requires that Container be in state [closed,hooked].

This E-GSM process model allows one to detect violations in the execution flow. For example, if we assumed that, once the truck reaches the inland terminal, the

container is detached from the truck without firstly inspecting the goods, stage
DetachContainer would become *outOfOrder* (being *DetachContainer.DFG*1 trig-
gered before *DetachContainer.PFG*1 becomes active), and stage *InspectGoods*
would become *skipped* (being the achievement of *InspectGoods.M*1 required for
*DetachContainer.PFG*1 to become active). Since these stages are not *onTime*,
the monitoring platform can compliance violation and, by inspecting this E-
GSM model, the we can determine that *DetachContainer* was executed before
InspectGoods.

5.1.4 Generating the E-GSM Artifact Lifecycle Model

To determine if the artifact embodied by the smart object that runs the platform
is correctly manipulated, the platform needs to know the expected lifecycle of the
artifact, i.e., the changes of state that are expected to occur during the execution of
the process. However, the E-GSM model derived in Section 5.1.3 does not contain
this information, and this limits the possibility of monitoring its compliance.

For example, still referring to the process presented in Section 3.1, let assume
that, after being filled in, the container overheats while it is still at the manufacturer's premises. Then it cools down before the truck arrives and, once the truck
reaches the manufacturer's premises, the process proceeds as planned. In this case,
no violation in the execution flow can be detected with the E-GSM process model,
since all activities are executed in the right order, even if the container was not
handled as expected.

Therefore, to provide this information to the monitoring platform, another E-
GSM model that captures the states that the artifact can assume during the execution of the process, and the admissible transitions from one of this states to
another one has to be defined. To automate the generation of this E-GSM artifact
lifecycle model, we defined a set of rules to automatically derive such a model from
an UML state chart diagram that captures the lifecycle of the artifact. The UML
state chart diagram, in turn, can be automatically derived from the BPMN process
diagram produced according to Section 5.1.2 by using the approach discussed by
Eid-Sabbagh et al. in [39]. Given an UML state chart diagram representing the
expected lifecycle of one artifact Ar, the following rules can be applied to automatically derive an E-GSM model of the lifecycle of Ar:

Rule An UML *state St* is translated into a *stage S''* with one *data flow guard*
$S''.DFG_1$, one *milestone* $S''.M_1$, and one *process flow guard* $S''.PFG_1$.
The ECA rule that defines $S''.DFG_1$ is triggered when a change in the state of Ar
occurs, and Ar enters the current state (which generates event Ar_e). It will only
be fired if the state assumed by Ar is the one indicated by St.
The ECA rule that defines $S''.M_1$ is triggered when a change in the state of Ar
occurs, and Ar leaves the previous state (which generates event Ar_l). It will only
be fired if the state assumed by Ar is different from the one indicated by St.
The boolean expression that defines $S''.PFG_1$ requires that at least one of the
stages that represent the UML *states* that directly precede St be opened (i.e.,

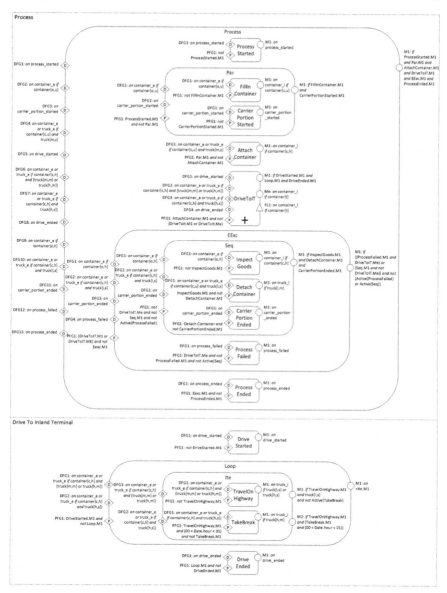

Fig. 5.9: E-GSM model representing the process from the container's viewpoint (top). For the sake of clarity, stages representing the **Drive to Inland Terminal** sub-process, i.e., stages inside `DriveToIT`, are shown separately (bottom).

active). If St is the initial state, $S''.PFG_1$ alternatively requires that no *stage* be active.   □

Rule A *stage* named $Error$, together with one one *data flow guard* $Error.DFG_1$, one *milestone* $Error.M_1$, and one *process flow guard* $Error.\ PFG_1$, is introduced to identify when Ar assumes a state not included in the UML state chart diagram. The ECA rule that defines $Error.DFG_1$ is triggered when a change in the state of Ar occurs, and Ar enters the current state (which generates event Ar_e). It will only be fired if none of the UML *states* St in the diagram indicates the state assumed by Ar.
The ECA rule that defines $Error.M_1$ is triggered when a change in the state of Ar occurs, and Ar leaves the previous state (which generates event Ar_l). It will only be fired if one of the UML *states* St in the diagram indicates the state assumed by Ar.
The boolean expression that defines $Error.PFG_1$ is never satisfied. This way, whenever Ar assumes one state not indicated in the expected lifecycle, $Error$ is opened and a violation in the lifecycle is detected.   □

Rule A *stage* named $Final$, together with one one *data flow guard* $Final.DFG_1$ and one *milestone* $Final.M_1$, is introduced to identify when Ar reaches a final state.
The ECA rule that defines $Final.DFG_1$ is fired when one stage S'' derived from an UML *state* representing a final state is opened.
The ECA rule that defines $Final.M_1$ is fired when none of the stages S'' derived from an UML *state* representing a final state is opened.

data flow guards and *milestones* allow the monitoring platform to keep track of the current state of Ar: when it is in a specific state, the corresponding *stage* are opened and the other ones, but $Final$, closed. When Ar changes state, the *data flow guard* attached to the *stage* that represents the new state is triggered, and the corresponding *stage* is opened. At the same time, the *milestone* attached to the *stage* that represents the previous state is achieved, and the corresponding *stage* closed. When an admissible state change occurs, both the *data flow guard* and the *process flow guard* of the *stage* that represents the new state are expected to be triggered. On the other hand, when a non admissible state change occurs, the condition on the *process flow guard* is not fulfilled, and therefore only the *data flow guard* is triggered.

To formalize these rules and automate the transformation, ATL was adopted again. In particular, Figure 5.11 shows a simplified version of the ATL translation rules to derive from the finite state machine obtained from [39] the corresponding E-GSM constructs[3]. Rule `state2substage` is responsible for transforming each stage in the source finite state machine into a *stage* in the target E-GSM lifecycle model. `state2substage` also produces, for each *stage*, one *process flow guard*, one *data flow guard*, and one *milestone*, that are attached to the corresponding *stage*.

---

[3] The complete version of these translation rules is available at https://bitbucket.
org/polimiisgroup/bpmn2egsm/src/206dd0270c4f32a7997d847356b0397fb283aac4/FSM2GSM/
fsm2egsm.atl.

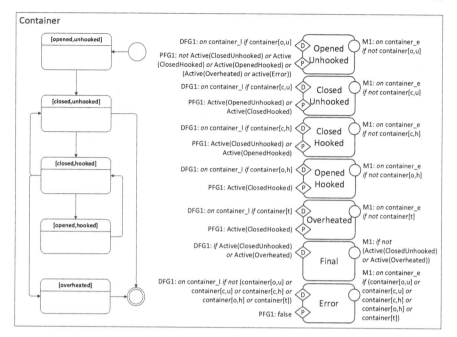

Fig. 5.10: Finite State Machine (left) and E-GSM model (right) representing the lifecycle of the container.

The ECA of the *data flow guard* and the *milestone* are triggered, respectively, when the artifact assumes the state, and when it assumes a different state. The boolean expression of the *process flow guard* is determined by the helper function `producePFGExpression`, which identifies the predecessor of the state and builds the expression accordingly. Finally, rule `fsm2stage` is responsible for producing the *Final* and *Error stages*, and their associated *process flow guard*, *data flow guard*, and *milestone*.

Example.

The left portion of Figure 5.10 shows an UML state chart diagram representing the lifecycle of the container, derived from the BPMN process diagram shown in the top portion of Figure 5.4 according to [39][4]. The container has a single initial state, [opened,unhooked]. This means that the process should start with an opened container that is not hooked to any means of transport. The two final states, [closed,unhooked] and [overheated], correspond to the process that terminates with

---

[4] Since we are not interested in identifying the activities responsible for causing the transition to fire, the labels on transitions are discarded.

```
1   rule fsm2stage {
2    from m: UML!StateMachine
3    to
4     tfdfg: EGSM!DataFlowGuardType (id <- 'Final_dfg1',
5       expression <- 'on '+m.name+'_l if ('+
6       'or'.concatStrings(m.getFinalStates()
7       -> collect(s | 'Active('+s.name+')'))+')'),
8     tfm: EGSM!MilestoneType (id <- 'Final_m1',
9       expression <- 'on '+m.name+'_e if not ('+
10      'or'.concatStrings(m.getFinalStates()
11      -> collect(s | 'Active('+s.name+')'))+')'),
12    tfs: EGSM!SubStageType (id <- 'Final',
13      dataFlowGuard <- OrderedSet{tfdfg},
14      milestone <- OrderedSet{tfm}),
15    tedfg: EGSM!DataFlowGuardType (id <- 'Error_dfg1',
16      expression <- 'on '+m.name+'_l if not ('+
17      'or'.concatStrings(m.getAllStates()
18      -> collect(s | m.name+'['+s.name+']' ))+')'),
19    tepfg: EGSM!DataFlowGuardType (id <-'Error_pfg1', expression <-'false'),
20    tem: EGSM!MilestoneType (id <- 'Error_m1',
21      expression <- 'on '+m.name+'_e if ('+
22      'or'.concatStrings(m.getAllStates()
23      -> collect(s | m.name+'['+s.name+']' ))+')'),
24    tes: EGSM!SubStageType (id <- 'Error',
25      dataFlowGuard <- OrderedSet{tedfg},
26      processFlowGuard <- OrderedSet{tepfg},
27      milestone <- OrderedSet{tem})}
28
29   rule state2substage {
30    from s: UML!State
31    to
32     tpfg: EGSM!ProcessFlowGuardType (id <- s.name+'_pfg1',
33       expression <- s.producePFGExpression()),
34     tdfg: EGSM!DataFlowGuardType (id <- s.name+'_dfg1',
35       expression <- 'on '+s.getParentStateMachine().name+'_l if '+
36       s.getParentStateMachine().name+'['+s.name+']' ),
37     tm: EGSM!MilestoneType (id <- s.name+'_m1',
38       expression <- 'on '+s.getParentStateMachine().name+'_e if not '+
39       s.getParentStateMachine().name+'['+s.name+']' ),
40     ts: EGSM!SubStageType (id <- s.name,
41       dataFlowGuard <- OrderedSet{tdfg},
42       milestone <- OrderedSet{tm},
43       processFlowGuard <- OrderedSet{tpfg})}
44
45   helper context UML!State def: producePFGExpression(): String =
46    'or'.concatStrings(self.getPredecessor() -> collect(t | 'Active('+t.name
47    +')') -> including(
48    if (self.isInitialState()) then
49      OrderedSet{ 'not ('+
50       'or'.concatStrings((self.getParentStateMachine().getAllStates()
51       -> excluding(self) -> collect(s | 'Active('+s.name+')'))
52       -> including('Active(Error)'))+')' }
53      else
54      OrderedSet{}
55      endif ) -> flatten());
```

Fig. 5.11: (Simplified) ATL translation rules to derive from a finite state machine the E-GSM lifecycle model a container that is either closed and unhooked from the truck, or that is overheated. Transitions among states allow the container to evolve.

The right portion of Figure 5.10, on the other hand, shows the E-GSM artifact lifecycle model obtained form the previously mentioned UML state chart diagram. *Final.DFG1* requires that stages *ClosedUnhooked* or *Overheated* be active, since stages [closed,hooked] and [overheated] are obtained from final states. For the same reason, *Final.M1* requires that both stages be not active. The *OpenedHooked.PFG1* requires that stage *ClosedHooked* be active, since the container is expected to enter state [opened,hooked] only if it exits state [closed,hooked], as there is only one transition between these two states in the state chart diagram. On the other hand, *ClosedUnhooked.PFG1* requires that either stage *OpenedUnhooked* or *ClosedHooked* be active, since the container can transition to [closed,unhooked] from either [opened,unhooked] or [closed,hooked]. Finally, *OpenedUnhooked.PFG1* requires that none of the other stages be active, since state [opened,unhooked] should be the initial state of the container.

5.1.5 Generating the Artifact-to-object Mapping Criteria

The E-GSM models generated in Section 5.1.3 and Section 5.1.4 allow the monitoring platform to monitor the process based on the state of the artifacts participating in that process. However, the E-GSM models do not specify which smart object will impersonate each artifact (e.g., the artifact Container is impersonated by the physical container having serial number "SN9876"). If a single artifact participates in the process, this information is not required, as the identity of that artifact coincides to the one of the smart object that runs the monitoring platform. On the other hand, if two or more artifacts participate in the process, their identity has to be notified to the monitoring platform. This way, the smart objects impersonating these artifacts will be able to exchange information on their state, thus allowing a complete monitoring of the process they participate in.

Since the identity of some smart objects may still be unknown once the process started, rules to communicate such an information at runtime are required. Additionally, once a smart object no longer participates in a process execution, it should no longer send notifications to the other smart objects participating in the same execution. To do so, the monitoring platform relies on events sent by the organizations when the process is run to determine when an artifact should be bound or unbound to a specific smart object. Therefore, criteria that map a specific event to the binding or the unbinding of an artifact with a smart object are required. Such mapping criteria are defined in a document which is kept separate from the E-GSM model. Such a choice allows us to decouple the process logic from the artifact instantiation logic, which significantly improves the scalability of the platform. This will be extensively discussed in Section 7.1. Starting from the BPMN process diagram produced according to Section 5.1.2, we defined the following set of rules to automatically derive the mapping criteria:

Rule A BPMN *data association Da* between a *generic, message, or signal start or intermediate catch event e* and an *output data object* referencing one artifact *Ar* with no data state is translated to a *mapping criterion Mb*.

Mb states that, whenever *e* occurs, *Ar* is bound to the smart object identified in the payload of *e*. Should the artifact be already bound to a different object, the new binding would replace the existing one.

For instance, when the event *Carrier portion started* occurs, *Truck* is bound to the physical truck whose license plate is specified in the payload of *Carrier portion started*.

Rule A BPMN *data association Da* between an *input data object* referencing one artifact *Ar* with no data state and a *generic, message, or signal end or intermediate throw event* is translated into a *mapping criterion Mu*.

Mu states that, whenever *e* occurs and *Ar* is bound to a smart object, it becomes unbound. If *Ar* is already unbound, no action is taken.

For instance, when the event *Carrier portion ended* or *Process failed* occurs, no truck is bound to *Truck*.

Also in this case we adopted ATL to formalize these rules and automate the transformation. Figure 5.12 shows the ATL translation rules to derive from the BPMN process diagram the artifact-to-object mapping criteria. Rule `dataobject2artifacts` is responsible for generating, for each artifact defined in the BPMN process diagram, as many binding and unbinding rules as the number of associated events. To do so, it relies on the helper functions `getBindingEvents` and `getUnbindingEvents` for identifying, respectively, the catch and throw events associated to the data objects representing the artifact. `dataobject2artifacts` also relies on the lazy rules `produceBinding` and `produceUnbinding` to produce the binding rules in the target model.

Example.

Figure 5.13 shows the artifact-to-object mapping criteria derived from the BPMN process diagrams of Figure 5.1.2. Since the identity of the truck is known only when event **Carrier portion started** occurs, a mapping criterion that binds **Container** is defined for that event. On the other hand, since the truck is no longer be related to the process after either event **Carrier portion ended** or *Process failed* occurs, mapping criteria that unbind **Container** are defined for these events. Similarly, the identity of the container is known only when event **Manufacturer portion started** occurs, and the truck is no longer be related to the process after either event **Terminal portion ended** or *Process failed* occurs. Therefore, mapping criteria that bind **Container** when **Manufacturer portion started**, and unbind it when **Terminal portion ended** or *Process failed* occur are defined.

```
1   helper context BPMN!DataObject def: getBindingEvents(): OrderedSet(BPMN!
        CatchEvent) =
2   BPMN!DataOutputAssociation.allInstances() -> select (t | t.targetRef =
3   self) -> collect (s | s.sourceRef) -> iterate(e; ret: OrderedSet(
4     BPMN!CatchEvent) = OrderedSet{} | ret -> including (
5     BPMN!CatchEvent.allInstances() -> select (ce | ce.dataOutputs
6     -> includesAll (e)))) -> flatten();
7
8   helper context BPMN!DataObject def: getUnbindingEvents(): OrderedSet(BPMN!
        ThrowEvent) =
9   BPMN!DataInputAssociation.allInstances() -> select (t | t.sourceRef
10  -> includes(self)) -> collect (s | s.targetRef)
11    -> iterate(e; ret: OrderedSet(BPMN!ThrowEvent) = OrderedSet{} |
12    ret -> including (BPMN!ThrowEvent.allInstances() -> select (ce |
13    ce.dataInputs -> includes (e)))) -> flatten();
14
15  lazy rule produceBinding {
16   from id: String
17   to
18    tb: DB!bindingEvent (
19      id <- id
20    )}
21
22  lazy rule produceUnbinding {
23   from id: String
24   to
25    tub: DB!unbindingEvent (
26      id <- id
27    )}
28
29  rule dataobject2artifact {
30   from sdo: BPMN!DataObject
31   to
32    ta: DB!artifact (
33      name <- sdo.name,
34      bindingEvent <- sdo.getBindingEvents() -> collect(be | thisModule.
              produceBinding(be.id)),
35      unbindingEvent <- sdo.getUnbindingEvents() -> collect(ube | thisModule.
              produceUnbinding(ube.id))
36    )}
```

Fig. 5.12: ATL translation rules to derive from a finite state machine the E-GSM lifecycle model

5.2 Proof of Correctness

The transformations from BPMN to E-GSM presented in Section 5.1.3 and Section 5.1.4 must be *correct*. With this term we intend that, given a BPMN model and a process execution trace, the trace deviates from the model if and only if the deviation can be detected with the derived E-GSM models. Since our monitoring platform is meant to monitor the process at runtime, we also require the E-GSM models to be prompt, i.e., the deviation can be detected as soon as it actually occurs. More specifically, let *(i)* $\mathcal{B}$ be the well-structured BPMN process diagram, obtained as discussed in Section 5.1.2, *(ii)* $\mathcal{G}_{\mathcal{B}}^P$ be the E-GSM process model representing the activities and their dependencies defined in $\mathcal{B}$, obtained as discussed in Section 5.1.3, *(iii)* $\mathcal{G}_{\mathcal{B}}^A$ be the E-GSM artifact lifecycle model encoding the admissible states and transition for one artifact Ar, obtained as discussed in Section 5.1.4,

```
1   <LocalArtifact name="Container"/>
2   <Mapping>
3     <Artifact name="Truck">
4       <BindingEvent id="Carrier_portion_started"/>
5       <UnbindingEvent id="Carrier_portion_ended"/>
6       <UnbindingEvent id="Process_failed"/>
7     </Artifact>
8   </Mapping>
9
10  <LocalArtifact name="Truck"/>
11  <Mapping>
12    <Artifact name="Container">
13      <BindingEvent id="Manufacturer_portion_started"/>
14      <UnbindingEvent id="Terminal_portion_ended"/>
15      <UnbindingEvent id="Process_failed"/>
16    </Artifact>
17  </Mapping>
```

Fig. 5.13: Artifact-to-object mapping criteria for the container (top) and the truck artifacts (bottom).

the translation is correct if and only if, for every (possibly partial) execution trace over the tasks and events of $\mathcal{B}$, the following assumptions hold:

- If the trace conforms to $\mathcal{B}$ (i.e., no deviation occurs) , then none of the stages of $\mathcal{G}_{\mathcal{B}}^P$ are *outOfOrder*. Conversely, if the trace contains a deviation, then such a deviation is promptly recognized by $\mathcal{G}_{\mathcal{B}}^P$, i.e., $\mathcal{G}_{\mathcal{B}}^P$ has at least one *outOfOrder*, *opened* stage when the deviation actually occurs. This is called the *execution flow alignment* between $\mathcal{G}_{\mathcal{B}}^P$ and $\mathcal{B}$.
- By projecting away data flow guards in $\mathcal{G}_{\mathcal{B}}^A$ (i.e., by keeping process flow guards only), $\mathcal{G}_{\mathcal{B}}^A$ has an evolution from one stage to another if and only if there exists a corresponding transition in the lifecycle of an artifact Ar that is induced by $\mathcal{B}$. This is called the *artifact lifecycle alignment* between $\mathcal{B}$ and $\mathcal{G}_{\mathcal{B}}^A$.

The formal proof showing that our translation mechanism is indeed correct is given in [87], and is also reported in Appendix B for convenience. In the following sections we present a high-level discussion of that proof.

5.2.1 Trace Conformance

Before proving that the translation preserves control flow and lifecycle alignment, we need to define what does it mean for a trace to *conform* to (and *deviate* from) the BPMN model $\mathcal{B}$, particularly taking into account the execution flow constraints defined in $\mathcal{B}$.

Typically, conformance is tackled by first transforming the process model of interest into a formal behavioral model, such as as a workflow net. Then, if all activities or events in a complete trace (i.e., a trace that contains a complete execution of the process) can be *replayed* in the same order as they are recorded in the trace, that is, starting from the initial state in the model, the ending state is reached, then the trace conforms to the model [116]. This procedure can be easily extended to partial traces. If a partial trace is equal to the prefix of a complete trace that conforms to the model, then that partial trace conforms to the model as well.

In our setting, we leverage the fact that $\mathcal{B}$ is well-structured. We also adopt an alternative definition of conformance that has three advantages: *(i)* it is modularly defined over the different types of blocks that may be employed to structure $\mathcal{B}$; *(ii)* it is also applicable to E-GSM, thus providing the basis for comparing $\mathcal{B}$ and $\mathcal{G}_{\mathcal{B}}^{P}$ in terms of execution flow alignment; *(iii)* it is fully compatible with the aforementioned definition of conformance.

To define conformance, we start by noting that no block is repeated twice in $\mathcal{B}$. This guarantees that activities and events are unambiguous, and at the same time ensures that no block directly or indirectly embeds itself. As a consequence, based on the relations among the sub-blocks composing $\mathcal{B}$, such blocks can be organized along a tree structure that has the top process block (i.e., the block that represents the complete process and includes all the other blocks) as the root node, and the single atomic activities and events as the leaves. We call such a tree the *process tree* of $\mathcal{B}$. The left portion of Figure 5.14 shows an excerpt of the process tree derived from the BPMN process diagram shown in Figure 5.4, relative to the Drive to inland terminal subprocess activity.

On top of this structure, a notion of *execution state* is introduced, so as to keep track of the currently active blocks, and of those that can be activated next. The initial execution state consists in the top process block being active, and the start event being ready to be activated next. When the start event occurs, the immediately consequent block becomes ready to be activated next. Given the current activation state, a new activation state is computed when the next execution step is performed, i.e., an event occurs, an activity is started, or an activity is completed. How the new state is computed depends on the specific types of the active blocks, and is done in two phases:

- **Check**. It is checked whether the execution step is accepted by $\mathcal{B}$ in the current activation state.
  The start of an activity or the occurrence of an event are accepted only if that activity or event can be activated next. Instead, the completion of an activity is accepted only if that activity is currently active. If the execution step is not accepted, then a deviation occurs.
- **Update**. If no deviation occurs (i.e., the execution step is accepted), then the current state of $\mathcal{B}$ is updated by deactivating active blocks, and by making new blocks active.

Both the check and the update phases depend on the semantics of the active blocks and of the ones that can be activated next. For example, a sequence block containing two tasks requires that *(i)* the first task can be activated as soon as the sequence block is activated, *(ii)* the second task can activated as soon as the first task is completed, and *(iii)* the sequence block is deactivated as soon as the second task is completed.

5.2.2 Execution Flow Alignment

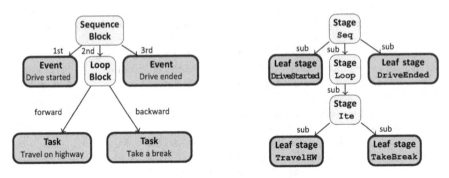

Fig. 5.14: Block structure of the Drive to inland terminal subprocess activity of the BPMN process diagram shown in Figure 5.4 (left), and of the stages in the E-GSM process model shown in Figure 5.9 derived from that activity (right).

Having described in Section 2.1.4 what are the requirements and implications for the acceptance of an execution step and the activation of a block, we can identify a direct, modular correspondence between $\mathcal{B}$ and $\mathcal{G}_{\mathcal{B}}^{P}$.

More specifically, the translation rules described in Section 5.1.3 guarantee that the process tree of $\mathcal{B}$ is modularly mirrored in $\mathcal{G}_{\mathcal{B}}^{P}$. In particular, the stage nesting relation of $\mathcal{G}_{\mathcal{B}}^{P}$ allows stages to be organized along a tree structure as well, which we name *stages tree*. Additionally, the structure of the stages tree reconstructs the process tree of $\mathcal{B}$ (possibly introducing single, intermediate stages to handle the semantics of the corresponding block in $\mathcal{B}$).

The right portion of Figure 5.14 shows the stage tree of the portion of the E-GSM process model shown in Figure 5.9, which is derived from the Drive to inland terminal subprocess activity. Note that every node in the process tree of that activity, which is shown in the left portion of Figure 5.14, is replicated in the stages tree. Additionally, the relations of ancestor, descendant or sibling among two nodes in the process tree are maintained by their correspondents in the stages tree.

Since the translation is modular with respect to the blocks of $\mathcal{B}$, and so is the notion of conformance sketched in Section 2.1.4, we then proceed by induction on the structure of the process tree. In particular, we show that, given an execution state s where all active blocks in $\mathcal{B}$ correspond to *opened, onTime* stages in $\mathcal{G}_{\mathcal{B}}^{P}$, and an execution step t:

- if t is not accepted by $\mathcal{B}$ in s (i.e., it is identified as a deviation), then the execution of t in s causes a stage of $\mathcal{G}_{\mathcal{B}}^{P}$ to become *outOfOrder*.
- if t is accepted by $\mathcal{B}$ in s, then the new execution state s' resulting from the execution of t in s is such that a block b is active in s' if and only if the corresponding stage in $\mathcal{G}_{\mathcal{B}}^{P}$ is *opened, onTime*.

The combination of these two properties implies that $\mathcal{G}_{\mathcal{B}}^{P}$ correctly monitors the control flow of $\mathcal{B}$, promptly detecting a deviation only when the currently processed execution step is indeed considered so by $\mathcal{B}$.

5.2.3 Artifact Lifecycle Alignment

Artifact lifecycle alignment amounts to check whether $\mathcal{G}_{\mathcal{B}}^{A}$ is constructed by properly considering the transitions in the state of an artifact Ar induced by $\mathcal{B}$. However, $\mathcal{G}_{\mathcal{B}}^{A}$ incorporates data flow guards that are not synthesized from $\mathcal{B}$, but are used to actually monitor the physical reality and obtain the state of the artifact accordingly. Hence, alignment is circumscribed to process flow guards.

It is immediate to see that, due to the way $\mathcal{G}_{\mathcal{B}}^{A}$ is derived, each state that can be assumed by Ar corresponds to a stage. Additionally, if Ar evolves as expected, only one of these stages will be *opened* throughout the execution of the process. We say that *artifact may change from state s_1 to state s_2 according to $\mathcal{G}_{\mathcal{B}}^{A}$ if $\mathcal{G}_{\mathcal{B}}^{A}$* foresees an execution step that: *(i)* is applicable when the stage corresponding to s_1 is *opened*, and *(ii)* causes that stage to close and to simultaneously open the stage corresponding to s_2.

Consequently, artifact lifecycle alignment consists in checking that, for every pair of states s_1 and s_2, Ar may change from state s_1 to state s_2 according to $\mathcal{G}_{\mathcal{B}}^{A}$ if and only if $\mathcal{B}$ foresees such a transition. This property, in turn, can be proven in two steps. In the first step, we rely on the correctness of the method proposed in [39], which encodes the state-transitions of the input BPMN model $\mathcal{B}$ into a corresponding state machine $\mathcal{M}$. In the second step, we reformulate the artifact lifecycle alignment by considering the explicit description provided by $\mathcal{M}$ instead of the implicit one obtained from $\mathcal{B}$. It is then straightforward to see that this reformulation: *(i)* faithfully encodes the behavior of $\mathcal{M}$; *(ii)* produces exactly $\mathcal{G}_{\mathcal{B}}^{A}$ from $\mathcal{B}$. Correctness of the artifact lifecycle then directly follows.

5.3 Summary

This chapter presented a method that, starting from standard BPMN collaboration diagrams, allows to semi-automatically derive the E-GSM models and the mapping criteria required for our artifact-driven monitoring platform to monitor the process. This way, organizations are not required to use E-GSM to model the process from scratch. Additionally, since BPMN is widely employed for documenting processes, it is likely that organizations already have a BPMN model of the process they intend to monitor. As such, the effort required to configure our artifact-driven monitoring platform becomes quite modest, making artifact-driven process monitoring applicable also to processes that are infrequently performed.

A disadvantage of this approach is the impossibility to fully exploit the declarative nature of GSM. Since the starting point is an imperative language (i.e.,

BPMN), purely declarative constraints cannot be defined with such a language. Consequently, the E-GSM models resulting from the application of this method will not include any of these constraints, although E-GSM allows to model them, as shown in Section 4.4. However, these constraints can be manually introduced by manually altering the resulting E-GSM models, thus combining the ease in reuse and in modeling of BPMN with the expressiveness of E-GSM.

Chapter 6
Assessing and Improving Process Monitorability

In Chapter 5, we presented a method to easily derive from a BPMN collaboration diagram the E-GSM process models and the criteria to map smart objects to artifacts. However, for the monitoring to be reliable, the smart objects participating in the process must autonomously determine all the states indicated in the process models. To do so, for every couple $\langle artifact, state \rangle$ indicated in a process model (e.g., by the data objects in the enriched BPMN collaboration diagram), the following requirements must be fulfilled:

- **MR1**: The smart objects embodying $artifact$ must be able to determine the physical properties of $artifact$.
- **MR2**: A relationship between $state$ and the physical properties of the smart objects embodying $artifact$ must exist.

To guarantee that these properties are fulfilled, for every smart object instantiating one of the artifact specified in the process model, the On-board Sensors Gateway and Events Processor components must be opportunely configured. To fulfill MR2, the Events Processor component must be instructed with rules to detect from the sensor data all the states of the artifact embodied by the smart object. To fulfill MR1, the On-board Sensors Gateway must interface with sensors capable to provide all the data required by the detection rules. If a smart object is unable to fulfill either MR1 or MR2, the monitoring platform running on that smart object is unable monitor when $artifact$ assumes $state$. Consequently, it cannot fully monitor the process.

Determining if these properties are fulfilled is far from trivial, especially when smart objects belonging to different organizations participate in the process. In fact, smart objects embodying the same artifact may have different sensors, sensor data may use different formats, or some smart object may lack sensors to measure a physical property (e.g., the speed). Additionally, the Events Processor may be implemented differently. For example, some smart objects may use a CEP, while others may use ad-hoc code. Therefore, state detection rules may be formalized using different languages.

To address these issues, in this section we present an ontology-based approach to assess and improve the *monitorability* of a process, i.e., given the capabilities of

© Springer Nature Switzerland AG 2019
G. Meroni: Artifact-Driven Business Process Monitoring, LNBIP 368, pp. 93–106, 2019
https://doi.org/10.1007/978-3-030-32412-4_6

the smart objects, to which extent it is possible to know when activities composing a process are executed [47]. Assessing the monitorability of a process model helps organizations to estimate the overall accuracy of a monitoring infrastructure, i.e., an artifact-driven platform together with the smart objects participating in the process, once the process is executed. Moreover, our approach can also be useful to improve process monitoring, as we are able to identify which activities are the most troublesome to monitor, and how to improve their monitorability by suggesting how to reconfigure the smart devices in terms of sensors and state detection rules.

6.1 Formalizing the Capabilities of the Smart Objects

To automatically determine to which extent smart objects can be used to monitor a process, their capabilities must be formalized. This allows smart objects that differ in the sensors and in the way the Events Processor component is implemented to be compared, and their fitness for monitoring to be evenly determined. To do so, ontologies [46] are adopted.

With respect to other data structures, like databases, ontologies guarantee better interoperability and reusability [126]. Each organization can autonomously maintain its own ontology, which describes the capabilities of the smart objects owned by that organization. Then, once two or more organization agree to participate in a process, their ontologies can be easily merged. This way, the monitorability of the process can be determined based on all the smart objects that may participate in that process, even though they belong to different organization. In addition, many ontologies to support the IoT are currently being developed and populated with information describing smart objects, sensors, and their capabilities, as discussed in Section 2.3.2. As such, organizations can reuse information already present in these ontologies, either by merging them or by linking concepts, instead of manually defining this information from scratch.

To formalize the capabilities of smart objects, we rely on two ontologies: *(i)* The *Smart Objects ontology*, that describes the hardware and software characteristics of the smart objects, and *(ii)* the *State Detection Rules ontology*, that formalizes the dependencies between the state of an artifact and the physical properties.

6.1.1 Smart Objects Ontology

Given a couple $\langle artifact, state \rangle$, to assess which smart objects fulfill MR1, the following characteristics should be taken into account:

- **MR1.1**: Which artifacts are embodied by the smart objects. This information is required to exclude from the assessment those smart objects that do not embody *artifact*.

- **MR1.2**: Who owns the smart objects. This way, it is possible to know the organization that should intervene in case a smart object is unfit to monitor a process.
- **MR1.3**: Which physical properties can be sensed by each smart object. This information is required to detect if a detection rule to detect *state* from those properties exists.
- **MR1.4**: Which units of measure are used to express the physical properties of a smart object. This way, mismatches between the format adopted by sensor data and the one required by detection rules can be taken into account.
- **MR1.5**: Which technology is adopted to implement the Events Processor. This way, detection rules that cannot be run by the Events Processor can be identified.

To formalize these characteristics, we chose to adopt and extend the FIESTA-IoT ontology [4], which is one of the most comprehensive ontologies for describing smart objects. By combining existing ontologies for the IoT, such as W3C SSN [32], M3 [49], and QU[1], FIESTA-IoT allows to describe the main function of a smart object (e.g., sensor, actuator, etc.), the capabilities of its sensing devices, and the physical properties that can be sensed from that smart object. Thus, it natively addresses MR1.3 and MR1.4.

FIESTA-IoT models hardware devices with a hierarchy of classes: *System*, which describes a generic hardware device, *Device*, a specialization of *System* that represents a hardware device dedicated to a specific purpose, and *Sensing device*, that indicates a hardware device dedicated to sense a physical property. *Sensing device* is also a specialization of the *Sensor* class, representing an instrument (not necessarily electronic) to sense a physical property. From now on, with the term sensor, we will always refer to a sensing device.

The *QuantityKind* class, linked to *Sensor* via the *hasQuantityKind* object property, indicates the physical property that is measured by a sensor. For example, to indicate that a speedometer measures the instantaneous speed, an individual *speedometer* of the *Sensor* class, an individual *speedinstantaneous* of the *QuantityKind* class, and an assertion of the *hasQuantityKind* object property among *speedometer* and *speedinstantaneous* have to be added to the ontology.

The *Unit* class, linked to *Sensor* via the *hasUnit* object property, indicates the unit of measure that is used by a sensor to represent a physical property. For example, to indicate that a speedometer expresses the speed in kilometers per hour, an individual *kilometerhour* of the *Unit* class, and an assertion of the *hasUnit* object property among *speedometer* and *kilometerhour* have to be added to the ontology.

Hardware devices (i.e., *System* elements) can also be aggregated to constitute an IoT platform, which is represented by the *ssn:Platform* class. A platform can be roaming (which is represented by the *iot-lite:isMobile* data property of a platform), or can be fixed (i.e., resides on a specific location).

However, FIESTA-IoT does not represent the relationships between the physical objects and the abstract artifact they impersonate, the owner, and the underlying technology, which are required to address MR1.1, MR1.2, and MR1.5, respectively.

---

[1] See http://purl.org/NET/ssnx/qu/qu

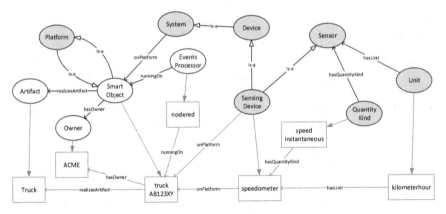

Fig. 6.1: Smart Objects ontology. Circles represent the classes (classes belonging to FIESTA-IoT are grayed out). Rectangles represent individuals. Grayed out lines represent property assertions among individuals.

Also, the concept of smart object, i.e., a physical object equipped with sensors, is not explicitly defined. Therefore, we extended FIESTA-IoT with the following concepts:

- **SmartObject** class. This concept is equivalent to the *ssn:Platform* class. Like a platform, a smart object is made of different components (i.e. devices), which may be sensors, actuators, computational or transmission modules.
  For example, a truck whose license plate is AB123XY is represented as an individual *truckAB123XY* of the *SmartObject* class.
- **Artifact** class. This concept represents the physical artifacts that can be instantiated by smart objects.
  For example, a generic truck is represented as an individual *truck* of the *Artifact* class.
- **realizesArtifact** object property. This property associates the *SmartObject* concept to the *Artifact* one. This way, it is possible to describe which smart objects instantiate an artifact, thus addressing MR1.1.
  For example, to specify that the truck whose license plate is AB123XY is a truck, the individual *truckAB123XY* is linked to the *truck* one by using *realizesArtifact*.
- **Owner** class. This concept represents the organizations who own the smart objects.
  For example, the carrier ACME is represented as an individual *acme* of the *Owner* class.
- **hasOwner** object property. This property associates the *SmartObject* concept to the *Owner* one. This way, it is possible to describe which smart objects are owned by which organization, thus addressing MR1.2.

For example, to specify that the carrier ACME owns the truck whose license plate is AB123XY, the individual *truckAB123XY* is linked to the *acme* one by using *hasOwner*.

- **EventsProcessor** class. This concept represents the technology used to implement the Events Processor component.
  For example, an Event Processor implemented with the Node-RED[2] flow engine is represented as an individual *nodered* of the *EventsProcessor* class.
- **runningOn** object property. This property associates the *EventsProcessor* concept to the *SmartObject* one. This way, it is possible to indicate which technology is used to implement the Events Processor component running on a smart object, thus addressing MR1.5.
  For example, to specify that truck whose license plate is AB123XY runs an Events Processor component implemented with Node-RED, the individual *nodered* is linked to the *truckAB123XY* one by using *runningOn*.

Figure 6.1 illustrates the (simplified) structure of the *Smart Objects ontology*, populated with some individuals.

6.1.2 State Detection Rules Ontology

Given a couple $\langle artifact, state \rangle$, to assess which smart objects fulfill MR2, the characteristics of the rules to detect when *artifact* assumes *states* given its physical should abstract the technology adopted to implement these rules. In particular, the following characteristics should be taken into account:

- **MR2.1**: Which states can be detected by a detection rule. This way, it is possible to exclude those rules that cannot detect *state*.
- **MR2.2**: Which physical properties are required by a detection rule. This information is required to determine if a smart object can provide these properties.
- **MR2.3**: Which units of measure are used to express the physical properties. This way, mismatches between the format adopted by sensor data and the one required by detection rules can be taken into account.
- **MR2.4**: Which technology is required to execute a detection rule. This way, it is possible to identify the smart objects that cannot execute a detection rule due to an incompatible Events Router.
- **MR2.5**: Whether or not it is possible to derive a physical property from other ones. This information turns to be useful when new state detection rules have to be modeled, and its implications will be discussed in detail in Section 6.4.

To formalize these characteristics, we adopted and extended the Physics Domain ontology introduced by Hachem et al. in [50]. The main advantage of this ontology is the possibility to define interdependencies among physical concepts, thus addressing MR2.5. To do so, formulas to convert a physical concept into another one (i.e., speed given space and time) are modeled in the ontology.

---

[2] See https://nodered.org.

In the Physics Domain ontology, such conversion formulas are modeled with the *Formula* class. To indicate which physical concepts are required for and derived from a conversion formula, the *Parameter* class is introduced and linked to *Formula* with, respectively, the *hasInput* and *hasOutput* object properties. To specify the physical concept of a parameter, *Parameter* is linked to the *QuantityKind* concept, imported from FIESTA-IoT, by the *hasConcept* object property. To specify the unit of measure of a parameter, *Parameter* is linked to the *Unit* concept, imported from FIESTA-IoT, by the *expressedInUnit* object property[3].

However, dependencies among physical concepts and states that can be assumed by the artifacts are not defined in the Physics Domain ontology, which are required to address MR2.1, MR2.2, MR2.3, and MR2.4. For this reason, in addition to the concepts presented in [50], we introduced the following elements in the State Detection Rules ontology:

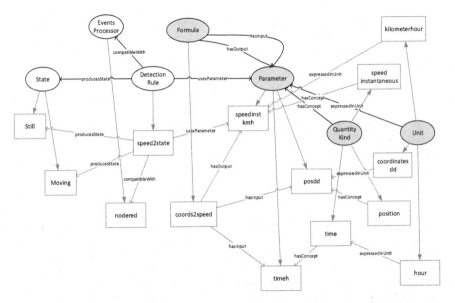

Fig. 6.2: State Detection Rules ontology. Circles represent the classes (classes belonging to the Physics Domain Ontology [50] are grayed out). Rectangles represent individuals. Grayed out lines represent property assertions among individuals.

- **State** class. This concept represents the discrete states that may be derived from sensor values.

  For example, the state *moving*, indicating that an artifact is moving, is represented as an individual of the *State* class.

---

[3] Actually, the Physics Domain ontology uses the name *hasUnit* to indicate such a property. However, *hasUnit* has already been used in FIESTA-IoT to indicate the relation between *Sensor* and *Unit*. Hence, to avoid ambiguity, we changed the name.

- **DetectionRule** class. This concept represents the state detection rules.
  For example, the rule *speed2state*, that determines if an artifact is still or moving based on its speed expressed in kilometers per hour, is represented as an individual of the *DetectionRule* class.
  Note that a state detection rule is different from a conversion formula: the former derives discrete states from physical concepts, while the latter converts sets of physical concepts into other physical concepts. Therefore, the *Formula* class cannot be used to express a state detection rule.
- **usesParameter** object property. This property associates the *DetectionRule* concept to the *Parameter* one. This way, it is possible to formalize which input data are required by the state detection rules, thus addressing MR2.2 and MR2.3.
  For example, to specify that the rule *speed2state* requires the speed of the artifact expressed in kilometers per hour as input parameter to operate, the individual *speedkmh*, referencing the physical concept of speed expressed in kilometers per hour, is linked to the *speed2state* one by using *usesParameter*.
- **producesState** object property. This property associates the *DetectionRule* concept to the *State* one. This way, it is possible to formalize which state detection rule can be used to derive a state, thus addressing MR2.1.
  For example, to specify that the state *moving* can be derived by using the rule *speed2state*, the individual *moving* is linked to the *speed2state* one by using *producesState*.
- **compatibleWith** object property. This property associates the *DetectionRule* concept to the *EventsProcessor* one, imported from the Smart Objects ontology. This way, it is possible to formalize which Event Processor components can run a state detection rule, thus addressing MR2.4.
  For example, to specify that rule *speed2state* is defined using Node-RED flow language, and as such can it can be run only on Events Processor implemented with Node-RED, the individual *speed2state* is linked to the *nodered* one by using *compatibleWith*.

6.2 Problem Setting

Guceglioglu et al. [47] measure the monitorability of a process as the ratio between the number of activities whose execution status (i.e., if they are started or ended) cannot be known and the total number of activities. By doing so, they expect the monitorability of each activity to be a binary value, i.e., either the activity can be fully monitored, or cannot be monitored at all. However, when an artifact-driven process monitoring platform is adopted, it may happen that an activity could be *partially* monitored. As we rely on the state of the artifacts to determine when activities are executed, and each execution may differ from the others in terms of smart objects embodying the artifacts, the monitorability of each activity may vary. In particular, when a smart object is unable to determine one state, the activites associated to that state cannot be monitored. Additionally, artifact-driven process monitoring distinguishes between the activation and the termination of activities. As such, it could be possible to know when an activity starts but not when it ends (or vice-versa), making that activity partially monitored.

To measure the process monitorability, we assume that the Smart Objects and the State Detection Rules ontologies contain the information about available smart objects, along with their capabilities. Thus, hereafter, $I = \{T^{SO}, T^{SDR}\}$ represents the monitoring infrastructure defined by the individuals of these ontologies.

Moreover, we assume the process to be modeled using a notation that associates artifacts and states to the activities composing the process, either as input or as output. In BPMN, artifacts and states can be represented with data objects, In E-GSM, this information can be defined with data flow guards and milestones. This way, the state held by the artifacts before and after the execution of an activity can be interpreted as the condition determining, respectively, the activation and termination of the activity.

More formally, we define a process model P as $P = \{A_i\}$, where:

$$A_i = \langle name, C_i^{start}, C_i^{stop} \rangle$$

is an activity defined by its name, the condition determining its activation, and the condition determining its termination.

For example, referring to the Drive to Inland Terminal subprocess activity in enriched BPMN collaboration diagram shown in Figure 5.2, we have two activities:

$$A_1 = \langle Travel\, on\, highway, C_1^{start}, C_1^{stop} \rangle$$
$$A_2 = \langle Take\, a\, break, C_2^{start}, C_2^{stop} \rangle$$

A condition determining the activation (termination) of an activity is defined as:

$$C_i^{start} = \{ARS_{i,j}\}, \quad C_i^{stop} = \{ARS_{i,k}\}$$

where $ARS_{i,j} = ARS_{i,k} = \langle artifact, \{state_l\} \rangle$ is the artifact, along with the states it assumes, that in the process model is associated to the i-th activity as input (output). A condition is true when all the associated artifacts are in the specified states.

For example, the conditions C_1^{start} and C_1^{stop}, determining, respectively, the activation and the termination of activity A_1 Travel on highway, are defined by:

$$C_1^{start} = \{\langle Container, \{closed, hooked\} \rangle, \langle Truck, \{highway, moving\} \rangle,$$

$$\langle Truck, \{manufacturer, moving\} \rangle\}\}$$
$$C_1^{stop} = \{\langle Truck, \{inlandterminal, still\} \rangle \langle Truck, \{highway, still\} \rangle\}$$

Thus, only when the container is closed and hooked, and the truck is moving either on the highway or at the manufacturer's premises (since $Truck$ is present twice with two different sets of states), then we can consider that the carrier started traveling on the highway. Similarly, when the truck is still either on the highway or at the inland terminal, the carrier stopped driving.

Based on this formulation, we want to assess the monitorability of a process model P with respect to the monitoring infrastructure I. Indeed, the ability of a smart object instantiating $Truck$ to detect if it is *moving* influences the monitorability of the activity Travel on highway. If no smart object has such ability, then not only this activity, but also Take a break can never be completely monitored, thus affecting the monitorability of the whole process.

6.3 Process Monitorability Assessment

To assess the monitorability of the process represented in a model P, we must first consider, for each couple $ARS_{i,j}$ in the process, to which extent is the monitoring infrastructure I suited to determine if $ARS_{i,j}.artifact$ assumes $ARS_{i,j}.state_l$.

We name this property, which can be computed querying the ontologies, *artifact state monitorability* $Mon^{ARS}(\langle artifact, \{state_l\}\rangle, I)$. To do so, we firstly identify the set of smart objects (i.e., all the individuals of the SmartObject class) able to instantiate the *artifact* :

$$SSO = \{so_n \mid so_n \in I.T^{SO} \wedge so_n.realizesArtifact = artifact\}$$

Then, we identify $\overline{SSO}^{state_l} \subseteq SSO$ containing only those smart devices so_n whose sensors can be used to detect $state_l$. For each so_n, the existence of at least one detection rule dr_d to derive $state_l$, whose input parameters par_p can all be provided by the sensors of so_n, is verified. To do so, for every par_p belonging to dr_d, the existence of a sensor device $sensDevice_s$ belonging to so_n whose measured quantity and unit of measure are the same as par_p, is verified. Additionally, dr_d must be compatible with the Events Processor ep_e running on so_n.

$$\exists dr_d \in I.T^{SDR} \mid dr_d.producesState = state_l \wedge$$
$$\exists ep_e \in I.T^{SO} \mid so_n \in ep_e.runningOn \wedge ep_e \in dr_d.compatibleWith \wedge$$
$$\forall par_p \in I.T^{SRD} \mid par_p \in dr_d.usesParameter \wedge$$
$$\exists sensDevice_s \in I.T^{SO} \mid sensDevice_s.onPlatform = so_n \wedge$$
$$par_p.hasConcept = sensDevice_s.hasQuantityKind \wedge$$
$$par_p.expressedInUnit = sensDevice_s.hasUnit \wedge$$

The cardinality of the set SSO and of the intersection of $\overline{SSO}^{state_l}$ for each state $state_l$ determines the artifact state monitorability, computed as:

$$Mon^{ARS}(\langle artifact, \{state_l\}\rangle, I) \to [0,1] = \frac{\left|\bigcap^{state_l} \overline{SSO}^{state_l}\right|}{|SSO|} \qquad (6.1)$$

The Mon^{ARS} is the basic building block for computing the process model monitorability as we need to check, for each activity belonging to the project, to which extent its activation and termination condition can be monitored, by computing the *condition monitorability* Mon^C. More formally, given $A_i = \{name, C_i^{start}, C_i^{stop}\} \in P$:

$$Mon^C(A_i.C_i^{start}, I) \to [0,1] =$$
$$\prod^{Ar} \frac{\sum^{ARS_{i,j} \in A_i.C_i^{start} \ni ARS_{i,j}.artifact=Ar} Mon^{ARS}(ARS_{i,j}, I)}{\left|ARS_{i,j} \in A_i.C_i^{start} \ni ARS_{i,j}.artifact = Ar\right|} \quad (6.2)$$

$$Mon^C(A_i.C_i^{stop}, I) \to [0,1] =$$
$$\prod^{Ar} \frac{\sum^{ARS_{i,k} \in A_i.C_i^{stop} \ni ARS_{i,k}.artifact=Ar} Mon^{ARS}(ARS_{i,k}, I)}{\left|ARS_{i,k} \in A_i.C_i^{stop} \ni ARS_{i,k}.artifact = Ar\right|} \quad (6.3)$$

Based on this definition, the monitorability of a condition $A_i.C_i^{start}$ determining the activation of A_i depends on the artifact state monitorability of the couples $ARS_{i,j}$ belonging to the condition. If the same artifact Ar is indicated for two or more couples $ARS_{i,j}$, then at least one of these couples is required to detect when A_i starts. On the other hand, if two couples $ARS_{i,j}$ indicate different artifacts, then they are all required to detect when A_i starts. Therefore, $Mon^C(A_i.C_i^{start}, I)$ is computed as the product, for each artifact, of the mean value of the monitorability of each couple $Mon^{ARS}(ARS_{i,j}, I)$ indicating the same artifact. The same considerations hold for the condition $A_i.C_i^{stop}$ determining the termination of A_i.

The monitorability of an activity is then defined by the monitorability of its activation and termination conditions, so that:

$$Mon^A(A_i, I) \to [0,1] =$$
$$\frac{1}{2} \cdot \left(Mon^C(A_i.C_i^{start}, I) + Mon^C(A_i.C_i^{stop}, I)\right) \quad (6.4)$$

We assume that the importance of determining when an activity starts or terminates is the same. It is worth noting that, if no start (termination) condition is put in the process model for a given activity, that activity is expected to be automated. Therefore, the contribution of Mon^C is 1, as the monitoring platform would always receive automatic notifications from the organizations.

Finally, we can define the process monitorability as:

$$Mon^P(P, I) \to [0,1] = \frac{\sum^{A_i \in P} Mon^A(A_i, I)}{|A_i \in P|} \quad (6.5)$$

Based on this definition, the process monitorability represents the average of the activity monitorability $Mon^A(A_i, I)$ computed for all the activities $\{A_i\}$ composing P. Following the same approach as [47], at this stage we do not consider the control flow as well as the probability of taking branches.

6.4 Process Monitorability Improvement

Besides quantitatively assessing the monitorability of a process, if the results do not meet the expectations of the organizations (i.e., the process monitorability value is too low), our ontology-based approach can also assist organizations to improve such results. Indeed, by analysing the monitorability at the different levels of granularities, the designer can identify which $\langle artifact, state \rangle$ couples contribute more to lowering the monitorability value. For each of these couples, different strategies can be followed: *(i)* The process model can be altered, *(ii)* new state detection rules can be introduced, or *(iii)* smart objects can be modified or replaced.

It is worth noting that, when a modification is introduced to improve the monitorability of a couple $\langle artifact, state \rangle$, that modification may also indirectly affect the other couples. Therefore, the contribution to the monitorability of a process should be computed from scratch whenever a new modification is introduced.

6.4.1 Process model improvement

The first strategy, i.e., modifying the process model, is particularly suited when a low monitorability value is caused by the impossibility to evaluate, given an $\langle artifact, state \rangle$ couple, when $artifact$ assumes $state$ (i.e., $Mon^{ARS}(\langle artifact, state \rangle, I) > 0$). This may occur when none of the smart objects embodying $artifact$ can provide sensor data to determine $state$, when no state detection rule to detect $state$ exists.

Assuming that current smart objects cannot be altered, new ones cannot be introduced, and new state detection rules cannot be defined (these cases will be discussed in the next sections), the monitorability of $\langle artifact, state \rangle$ can be improved (i) by finding, for the same artifact, a monitorable state, i.e., $\langle artifact, state' \rangle$, or (ii) by finding a different artifact able to monitor that state, i.e., $\langle artifact', state \rangle$.

In the former case, by querying the ontologies, we can obtain the alternative states $state'$, for the same artifact $artifact$ already specified in the model, for which the infrastructure can ensure a better monitorability:

$$state' \in I.T^{SDR} \mid Mon^{ARS}(\langle artifact, state' \rangle, I) > 0$$

For example, if $Mon^{ARS}(\langle truck, moving \rangle, I) = 0$ and $Mon^{ARS}(\langle truck, accelerating \rangle, I) > 0$, the ontologies suggest to replace $\langle truck, moving \rangle$ with $\langle truck, accelerating \rangle$.

In the latter case, the ontologies suggest to use another artifact $artifact'$ for which the sensors are able to return the occurrence of the same state $state$:

$$artifact' \in I.T^{SO} \mid Mon^{ARS}(\langle artifact', state \rangle, I) > 0$$

For instance, if the $\langle container, hooked \rangle$ cannot be monitored, as no container has the related sensors, it could happen that the truck has these sensors. Thus, replacing that couple with $\langle truck, hooked \rangle$ would improve the monitorability.

In any case, organizations are responsible for deciding which modifications can be applied to the process model without changing its behavior. Indeed, being a domain-dependent problem, our approach can only provide suggestions.

6.4.2 State detection rules improvement

Another possible improvement strategy consists in the introduction of new state detection rules. In particular, given an $\langle artifact, state \rangle$ couple for which more than one smart object instantiating $artifact$ is not able to monitor $state$, a new state detection rule, whose input parameters can all be provided by the smart object, can be introduced. This is only possible when a cause-effect relation among the values coming from the sensors of the smart object and $state$ exists.

In some cases, a state detection rule dr_d to infer $state$ may already exists, but it may not be compatible with the Events Processor ep_e running on a smart object so_n. This causes so_n to be unable to monitor $state$. However, this issue can be solved by simply adding a new detection rule dr'_d that behaves exactly as dr_d, but that has been rewritten to be compatible with ep_e.

For example, suppose that the Events Processor of truck $truckCD456WZ$ is implemented with WSO2CEP[4]. Yet, the state detection rule $speed2state$, to infer when the truck is $moving$, is formalized as a Node-RED flow. This causes this truck to be unable to detect when it is moving. However, a new state detection rule compatible with the Events Processor of truck $truckCD456WZ$ can be derived by rewriting $speed2state$ using WSO2 CEP-compatible language.

By querying the ontologies in the same way as when identifying $\overline{SSO}^{state}$, except for removing the constraint on $dr_d.compatibleWith$, we can identify a new set of smart objects $\overline{SSO}^{state}_{future}$. Then, by computing the difference between $\overline{SSO}^{state}_{future}$ and $\overline{SSO}^{state}$ we can identify which detection rules can be rewritten, and how the introduction of a new one would impact on the monitorability.

Another case concerns a state detection rule dr_d that, to infer $state$, may require input parameters expressed with an unit of measurement different that the one used by the sensors of a smart object so_n. This causes so_n to be unable to monitor $state$. For this case, adding a new detection rule dr'_d identical to dr_d except for the presence of a conversion formula solves this issue.

For example, suppose that the speedometer of the truck $truckEF789AB$ expresses the speed in miles per hour. Yet, the state detection rule $speed2state$, to infer when the truck is $moving$, requires as input the speed expressed in kilometers per hour. Even though $speed2state$ cannot be used to detect when the truck is mov-

---

[4] See http://wso2.com/products/complex-event-processor/.

ing, a new state detection rule can be easily derived from *speed2state* by simply converting the unit of measure of the input parameter.

In this case, set $\overline{SSO}_{future}^{state}$ can be determined as above, except for removing the constraint on $par_p.expressedInUnit$.

A third case concerns a state detection rule requiring one parameter that cannot be provided by the smart objects. Yet, that parameter can be derived from the ones provided by the smart objects. For example, suppose that the truck *truckGH135CD* is not equipped with a speedometer, yet it has a GPS transponder and an internal timer. In this case, the state detection rule *speed2state* cannot be used, as it requires the speed as input. However, knowing the instantaneous position of the truck from the GPS transponder, and the current date and time from the timer, it is possible to derive the speed from these physical concepts by applying the *timecoords2speed* conversion formula. Therefore, *speed2state* and *timecoords2speed* can be used to derive a new state detection rule that uses these physical concepts as input parameters, instead of the speed.

Also in this case, set $\overline{SSO}_{future}^{state}$ can provide insights on which detection rules and conversion formulas to use. In particular, the ontology should be queried in the same way as before, except for requiring every parameter of dr_d to be either provided by a smart object so_s, or by a conversion formula cf_f, whose input parameters are either provided by so_s or, recursively, by another conversion formula cf'_f.

6.4.3 Infrastructure improvement

A third improvement strategy can be adopted when, given an $\langle artifact, state \rangle$ couple, the smart objects cannot provide the required information to derive *state*, neither directly nor indirectly, i.e., the difference between $\overline{SSO}_{future}^{state}$ and $\overline{SSO}^{state}$ is an empty set. In this case, by querying the ontology, we can obtain information on how to alter the existing smart objects. In particular, for every smart object $so_n \in SSO \backslash \overline{SSO}_{future}^{state}$, for every detection rule dr_d such that $dr_d.producesState = state$, the physical properties required by the input parameters par_p of dr_d, minus the ones already provided by so_n, are returned.

This way, so_n can either be replaced, or altered by adding new sensors, such that all par_p of at least one dr_d can be provided. The applicability of this strategy depends on the number of smart objects to alter or replace, and on their accessibility, cost, and ownership.

For example, suppose that the truck *truckHI579TY* is only equipped with an indoor humidity sensor. In this case, no correlation exists between the truck being on the move and the indoor humidity. Therefore, no detection rule dr_d uses humidity data to infer if the truck is *moving*. So, to infer that state, either additional sensors have to be installed, or the truck has to be replaced.

6.5 Summary

This chapter presented an ontology-based approach to determine if a monitoring infrastructure can reliably monitor a business process. In particular, two ontologies were used to formalize the characteristics of smart objects, their sensors, and the rules to derive from sensor data the state of the artifacts. The monitorability metric was adopted to quantify to which extent a process can be monitored. It was then shown that, by opportunely querying the ontologies, it is possible to automatically compute the monitorability of a proces based on the capabilities of the smart objects. Finally, ontologies were used to provide suggestions to the process modeler on how to improve the monitoring platform.

Currently, the approach is being applied to logistics and validated by domain experts in the context of the Italian project ITS Italy 2020. Additionally, preliminary development of a tool supporting the approach has also been started. The ontologies supporting this approach have been implemented with Protégé [96], and are publicly available at http://purl.org/polimi/martifact/sosdr. The usage of SPARQL Protocol and RDF Query Language (SPARQL) [105] queries to automatically query the ontologies, compute the monitorability, and identify good candidates among the possible modifications is currently being investigated.

Chapter 7
Implementing and Evaluating Artifact-driven Process Monitoring

To evaluate the applicability of our artifact-driven monitoring approach, we developed a prototype implementing the reference architecture described in Section 3.3 using the Node.js runtime environment[1]. Node.js eases the execution of hardware and operating system-agnostic JavaScript code, is available for most hardware and software platforms, and is optimized for systems low on CPU power and RAM. This way, we were able to run the resulting platform prototype on several resource-constrained SBCs, such as the Intel Galileo[2] and the RaspBerry Pi[3].

This prototype was then tested within both a simulated environment and a field environment. For the simulated environment, truck shipments data provided by an European logistics company were used. For the field environment, realtime data from a GPS receiver directly attached to the SBC running the software component were used.

7.1 SMARTifact: an Artifact-driven Monitoring Platform

Figure 3.3 shows the architecture of SMARTifact, our prototype of artifact-driven monitoring platform. The **E-GSM Engine**, the component responsible for monitoring the execution of the process, is the heart of SMARTifact and is directly run on top of each smart object. This component takes as input the E-GSM model of the process and, based on the state of the artifacts participating in the process execution *(i)* keeps track of which activities are ongoing, *(ii)* keeps track of the current state of the artifact embodied by the smart object, *(iii)* identifies which activities are incorrectly executed, *(iv)* identifies which activities violate the execution flow, *(v)* identifies if the artifact incorrectly transitions from one state to another one. Instead of extending one of the few GSM engines, such as Barcelona [54] we chose to implement our E-GSM Engine from scratch[4]. This allowed us to optimize the E-

---

[1] See https://nodejs.org.

[2] See https://software.intel.com/en-us/iot/hardware/galileo.

[3] See https://www.raspberrypi.org.

[4] Source code at https://bitbucket.org/polimiisgroup/egsmengine.

© Springer Nature Switzerland AG 2019
G. Meroni: Artifact-Driven Business Process Monitoring, LNBIP 368, pp. 107–120, 2019
https://doi.org/10.1007/978-3-030-32412-4_7

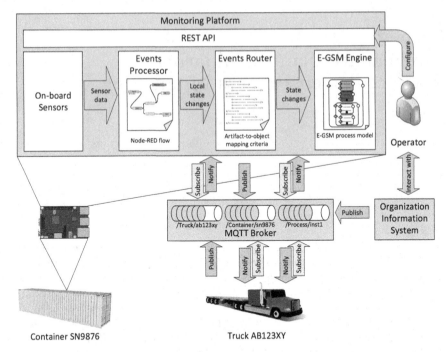

Fig. 7.1: Architecture of the artifact-driven monitoring platform prototype.

GSM Engine for monitoring the process (rather than for automating the execution) and for running on resource-constrained devices, such as SBCs.

The **Onboard Sensors Gateway**, i.e., the component responsible for collecting sensor data and determining the physical characteristics of the smart object, is the only component dependent on the hardware of the smart object. To implement such a module, we relied on the Node.js modules GPS.js[5] to extract the information provided by the GPS receiver. This module also produces a log of the sensor data, which can be used to identify possible issues in the monitoring platform.

To implement the **Events Processor**, i.e., the component responsible for inferring the state of the artifact embodied by the smart object from sensor data, we adopted the Node-RED[6] flow engine. Node-RED relies on a graphical flow-based language to process sensor data. This way, very limited programming skills are required to configure and customize this module. Consequently, the rules that determine how the state of the artifact is inferred can be directly defined by domain experts.

To dynamically exchange information on the state of the artifacts with the other smart objects, and to receive notifications from the information systems of

---

[5] See https://www.npmjs.com/package/gps.

[6] See https://nodered.org.

the organizations participating in the same process execution, the *Events Router*[7] component interacts with a **Message Queue Telemetry Transport (MQTT) Broker**. MQTT[8] is a queue-based publish/subscribe protocol, which is especially suited for applications where computing power and bandwidth are constrained.

The MQTT Broker contains as many topics (i.e., queues) as the smart objects participating in the process. Each of these topics adheres to the following naming convention: /{artifact_type}/{object_id}, where artifact_type is the artifact embodied by the smart object (e.g., a container or a truck), and object_id is the unique identifier of the smart object (e.g., the serial number of the container or the license plate of the truck). Whenever a smart object changes its state, it publishes the updated state on its own topic.

The MQTT Broker also contains as many topics as the process executions that are currently running. Each of these topics adheres to the following naming convention: /{process_name}/{instance_id}, where process_name is the name of process model to be monitored (i.e., the shipment process described in Section 3.1, henceforth FirstMile), and instance_id is the unique identifier of the process instance (i.e., the actual execution of the process) that is being run. These topics are used by the information systems of the organizations to send events related to the running processes, but not related to the state of the artifacts (i.e., when a subprocess starts or ends).

The Events Router is notified by the Events Processor whenever the artifact embodied by the local smart object, i.e., the smart object that runs these components, changes its state. When such a change of state occurs, the Events Router publishes it to the topic /{artifact_type} /{object_id}, where object_id and artifact_type are the unique identifier and the artifact instantiated by the local smart object, and forwards it to the E-GSM Engine.

By using the artifact-to-object mapping criteria, such as the ones produced according to 5.1.5, the Events Router can be notified on which other smarts objects, besides the local one, take part in the same process execution even after the process started. To do so, the Events Router subscribes to the topic related to the process execution in which the local smart object takes part (e.g., /MtoCProcess/inst1). Whenever a new event is published on that topic (e.g., process_started), the Events Router checks if a mapping criterion is defined for that event. If no mapping criterion exists, the Events Router simply forwards the event to the E-GSM Engine. If a binding criterion exists, the Events Router subscribes to the topic /{artifact_type}/{object_id}, where object_id is the remote smart object specified in the payload of the event (e.g., /Container /sn9876). From that point on, whenever a new change of state is published in /{artifact_type}/{object_id}, the Events Router forwards it to the E-GSM Engine of the local smart object. For example, if the truck having license plate AB123XY publishes on /Truck/AB123XY that its state changed to heathrow,moving, the Events Router of the container will notify that Truck is in heathrow,moving, together with the raising of Truck_e and Truck_l events, to the E-GSM Engine of the container. If an unbinding criterion exists, the Events Router unsubscribes to

---

[7] Source code at https://bitbucket.org/polimiisgroup/eventsrouter.

[8] See http://mqtt.org.

the topic /{artifact_type}/{object_id}, where object_id is the remote smart object specified in the payload of the event.

Note that, by keeping the binding logic separate from the process logic, the E-GSM Engine receives only events coming from those smart objects that are bound to the same process execution. This way, the scalability of the E-GSM Engine is affected only by the number of smart objects effectively participating in the same process execution. Finally, the **Representational State Transfer (REST) API** [113] offers an interface for the organization that owns the smart object to interact with it. It allows *(i)* the Events Processor to be configured with Node-RED flows, *(ii)* the E-GSM Engine to be provided with the E-GSM model, *(iii)* the Events Router to be instructed with the artifact-to-object mapping criteria, and *(iv)* the organization to determine if the process is correctly executed. In addition to that, it is responsible for the management of the communication channel between the organizations and the smart object: Whenever a new process execution takes place, an unique identifier *instance_id* of that process execution is agreed among the organizations participating in that process execution. Once the REST API is notified of *instance_id*, it instructs the MQTT Broker to subscribe to the /{process_name}/{instance_id} topic.

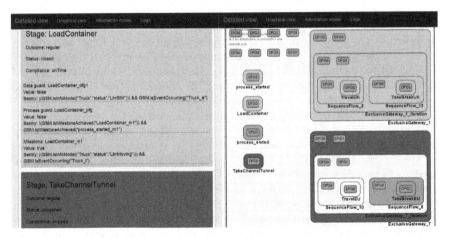

Fig. 7.2: Screenshot showing the graphical interface of SMARTifact.

All these modules were implemented in Node.js and, with the exception of the MQTT broker, were deployed on an Intel Galileo SBC. Events coming from the organizations' information systems were simulated by manually creating and directly publishing them on the /{process_name} /{instance_id} topic with the MQTT.fx tool[9]. Also, to visually inspect the process instance being monitored, a web-based graphical interface was developed, which is shown in Figure 7.2. The *Graphical view* (right) shows the stages composing the process and their nesting

---

[9] See http://mqttfx.jensd.de

relationship, whereas the *Detailed View* (left) shows the current state of each stage, along with details on its Data Flow Guards, Process Flow Guards, Milestones, and Fault Loggers. To easily identify if the process is not being correctly executed and, in case so, which portions of the process are affected by a violation, different colors are assigned to stages depending on their state: Light gray for *unopened, regular* stages. Yellow for *opened, onTime, regular* stages. Green for *closed, onTime, regular* stages. Dark gray for *skipped* stages. Red for *outOfOrder* or *faulty* stages. The *Information Model* view (not depicted) shows the contents of the Information Model. Finally, the *Logs* view (not depicted) lists all the events determining a change in the state of the stages, along with the affected stages, and a timestamp recording when the event took place.

The response time of SMARTifact, i.e. the time taken by the platform to process sensor data, depends on the throughput of the sensor data, and on the complexity of the discretization rules and of the process model. In our experimental setting we observed that, with medium-complexity discretization rules and process models, the response time stays under 100 milliseconds as long as less than three sensor reads per second are performed. We also observed that, for more than 20 sensor reads per second, SMARTifact saturates and it is no longer able to keep up with the sensor data (i.e. processing time grows based on the total number of sensor reads performed). This behavior is caused by SMARTifact running out of memory, thus relying on paging to flash storage (i.e., memory card). In fact, being the Intel Galileo equipped with only 256 MB of RAM, when SMARTifact is deployed it pre-allocates all the available memory. Nevertheless, we find these performance results to be adequate for the vast majority of the applications.

7.2 Simulated Environment

To demonstrate the applicability and efficacy of artifact-driven process monitoring in determining when activities are carried out without human interaction, we have conducted an experiment with truck shipments data provided by a European logistics company. [10] This provided material consisted of *(i)* a dataset with the registered positions and speed of trucks involved in the shipments, captured by on-board GPS systems and henceforth indicated as *GPS log*, and *(ii)* a dataset indicating the shipments' activities start and completion times, manually triggered by the truck drivers and hereinafter denoted as *activity log*. More in detail, each entry in the GPS log represents a sensor read. For each entry, a timestamp indicating when sensor data were collected, the identity of the truck, the route followed by the truck and an identifier of the process execution are recorded. In the activity log, on the other hand, each entry represents a notification sent by the truck driver. For each entry, a timestamp indicating when the notification was sent, the involved activity, the type of notification (start or stop), the route followed by the truck and an identifier of the process execution are recorded.

---

[10] The (anonymized) dataset is available at `http://purl.org/polimi/martifact/ logisticsds-anon` (password: GM-CDC-JM-dataset).

Shipments within continental Europe are organized as follows: At first, the container is attached to the truck. Then, the truck starts traveling until either a break is taken or the destination is reached. The alternation of traveling hours with breaks forms a loop: Once the break ends, the truck starts traveling again. Shipments that connect continental Europe with the United Kingdom are slightly different: After attaching the container, the truck alternates traveling with breaks until the entrance of the Channel tunnel is reached. Once there, the Channel tunnel is taken and, after that, the truck starts traveling again, possibly taking breaks, until it reaches the destination.

The experiment consisted in replaying the GPS log within our platform and checking whether the start and completion events detected by our platform matched with the manually inserted information in the activity log. This way, we were able to compare our artifact-driven monitoring platform with a traditional one relying on notifications sent by human operators.

We focused on routes that connect the headquarters of the logistics company, located in Amsterdam (AMS), to four European airports: London Heathrow (LHR), Brussels (BRU), Paris Charles de Gaulle (CDG), and Frankfurt (FRA). For every route, we considered both inbound and outbound routes from/to Amsterdam. The GPS log and the activity log contained 19 966 and 815 entries, respectively, distributed over 77 shipments. The reported shipments took on average 533 minutes, ranging from less than 3 to more than 27 hours.

The experiment was carried out as follows:

1. We analyzed the activity log and, for each route, we derived a BPMN collaboration diagram representing the process carried out for each route.
2. By inspecting the GPS log, we identified all the possible discrete states that each truck can assume. Using Node-RED, we also defined the rules to derive the state of the truck by replaying the GPS log, which are shown in Figure 7.3. In particular, for each entry in the GPS log, we embedded its timestamp into the corresponding discretized state. We also propagated this discretized state only if it was different from the previous one (i.e., if two entries in the GPS log were discretized into the same state, only the state related to the first occurrence was propagated).

   For example, for the LHR-AMS route we identified 12 discrete states: [heathrow,moving] and [heathrow,still] when the truck is, respectively, moving and not moving nearby London Heathrow airport. [amsterdam,moving] and [amsterdam,still] when the truck is nearby Amsterdam. [cheriton,moving] and [cheriton,still] when the truck is nearby the entrance of the Channel tunnel in the UK. [coquelles, moving] and [coquelles,still] when the truck is nearby the entrance of the Channel tunnel in France. [highwayUK,moving] and [highwayUK,still] when the truck is traveling along a highway in the UK. [highwayEU,moving] and [highwayEU,still] when the truck is traveling along a highway in the European continent.
3. Following the method presented in Chapter 5, we enriched each BPMN collaboration diagram with artifacts representing the truck and its states. Then, we derived the BPMN process diagrams and, from these diagrams, we generated the E-GSM models and the artifact-to-object mapping criteria.

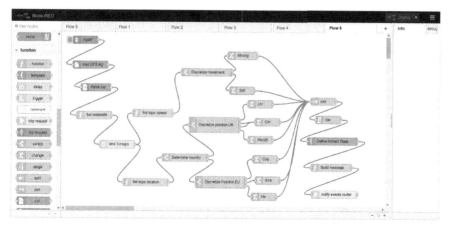

Fig. 7.3: Screenshot showing the rules to derive the state of the truck from the GPS log modeled with Node-RED.

For example, Figure 7.4 shows the enriched BPMN collaboration diagram of the process carried out for the LHR-AMS route.

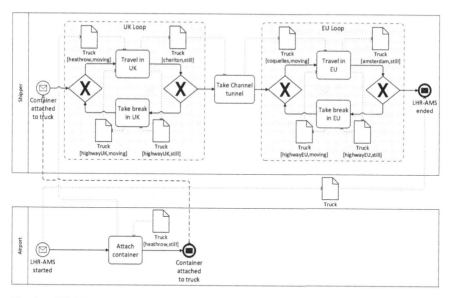

Fig. 7.4: BPMN collaboration diagram of the process carried out for the LHR-AMS route.

4. We deployed SMARTifact on an Intel Galileo SBC. We then instructed such a platform with the rules, process models and mapping criteria derived in the previous steps.

5. Instead of using data coming from sensors, we replayed the GPS log and relied on our platform to monitor the associated process.

6. Finally, we compared the results of the monitoring platform with the activity log. In particular, we verified that, whenever a stage changed its execution status (e.g., stage Take Channel tunnel transitioned from *unopened* to *opened*), an entry in the activity log indicated that the corresponding activity existed (e.g., the activity Take Channel tunnel was started).

If no entry in the activity log was related to such a stage, the corresponding process execution was flagged as *incompletely monitored*. Otherwise, we computed the difference between the timestamp associated to the last discretized state causing the stage to change its status (e.g., the timestamp associated to Truck[cheriton,still]) and the timestamp associated to the corresponding entry in the activity log (e.g., the timestamp associated to the entry indicating that activity Take Channel tunnel was started). We named such time difference as *detection delay*[11].

Also, we verified that executions that were known to be correctly performed did not contain any *skipped, outOfOrder*, or *faulty* stage. If one of such stage was present, the execution was flagged as *incorrectly monitored*. Likewise, we verified that executions that were known to be incorrectly performed were identified as such, that is, one or more stages were *skipped, outOfOrder*, or *faulty*, depending on the violation. If such stages were not present, the execution was flagged as *incorrectly monitored* as well.

Table 7.1: Results of the experiment on a simulated environment.

Shipment	AMS-LHR	LHR-AMS	AMS-BRU	BRU-AMS	AMS-CDG	CDG-AMS	AMS-FRA	FRA-AMS	Global
Instances	12	15	9	11	8	10	4	8	77
Median duration [min]	806.28	720.05	306.67	256.30	813.48	483.69	481.32	396.30	533.01
Min. duration [min]	338.47	138.02	153.00	159.62	387.57	353.00	396.10	279.32	138.02
Max. duration [min]	1328.56	1622.03	519.12	388.30	1583.52	723.25	567.47	357.32	1622.03
Correctness [%]	91.67%	100.00 %	100.00 %	90.91%	100.00 %	100.00 %	75.00%	87.50%	93.13%
Completeness [%]	58.33%	53.33%	77.78%	90.91%	87.50%	60.00%	100.00 %	62.50%	73.79%
Median detection delay [min]	2.73	−0.50	5.33	1.09	14.79	0.80	7.10	2.44	4.22
Median absolute d. delay [min]	12.53	4.57	7.10	5.17	16.57	4.18	8.87	4.88	7.98

Table 7.1 shows the results of this experiment. With the term *correctness* we indicate the difference between the incorrectly monitored executions and the total number of executions, divided by the total number of executions. Similarly, with the

---

[11] To compute the detection delay, we did not take into consideration the response time of the monitoring platform. As we limited the number of entries in the GPS log that were replayed to 3 per second, the delay introduced by the response time (that is on average in the order of magnitude of a tenth of a second) can be ignored.

term *completeness* we indicate the difference between the incompletely monitored executions and the total number of executions, divided by the total number of executions.

The monitoring platform was able to correctly determine the actual execution of a process for 93.13% of the total instances. For the remaining 6.87%, the platform had issues in determining when the container was attached to the truck. For example, during one shipment of the BRU-AMS route, activity Attach container was not identified as completed, even though it was. This was caused by the limited information that could be used to determine the state of trucks: As previously stated, the GPS log contained only information on the speed and position of the trucks. Thus, the anomalous slow progression of the trucks due to congested roads and traffic jams at the airports caused the misinterpretation of their state.

Moreover, the monitoring platform detected activities to be started or ended more often than what had been recorded in the activity log. Thus, the completeness amounted to 73.79%. Whether the missing entries in the activity log were due to an omission of the driver, or rather due to a wrong detection of the system, is debatable and needs further investigation. However, e.g., whenever the monitoring service notified that activity Travel in EU was ended, and no notification was sent by the truck driver, we inspected the GPS log and noticed that the truck had reached Europe and its speed had amounted to zero for more than a quarter of an hour, which suggests the first hypothesis to be more likely.

To assess the time gain for the detection of the status changes in the process, we relied on the detection delay. On average, the median of the detection delay amounted to 4.22 minutes (7.98 considering the absolute values of the delay), which is negligible for processes that last on average 533 minutes.

7.3 Field Evaluation

The experiment discussed in Section 7.2 was carried out by replaying a log containing sensor data. To demonstrate that artifact-driven process monitoring can also deal with live sensor data, a field evaluation of SMARTifact was performed. For this evaluation, we chose to monitor the processes the author of this book perform in order to reach his office from his house, and vice-versa. We name such processes VA-MI and MI-VA, respectively.

The VA-MI process is organized as follows. At first, the author walks from his house, located in Varese, to the Varese FS and Varese Nord railway stations (these stations are close to each other). Once there, he takes a train to Milan. From September till July he takes the route from Varese FS to Milano Garibaldi railway station. In August, on the other hand, he takes the route from Varese Nord to Milano Cadorna railway station, as this route is cheaper than the previous one only during that month. Once in Milan, the author takes the subway to either Piola or Lambrate subway stations and, once he reaches these stations, he finally walks to his office.

The MI-VA process is similarly organized. At first, the author leaves his office and walks to either Piola or Lambrate subway stations. Once there, if it is the 6th of the month, he takes the subway to Milano Garibaldi railway station, and then a train to Varese FS railway station. Otherwise, he takes the subway to Milano Cadorna, and then a train to Varese Nord railway station. Once in Varese, he finally reaches his house by foot.

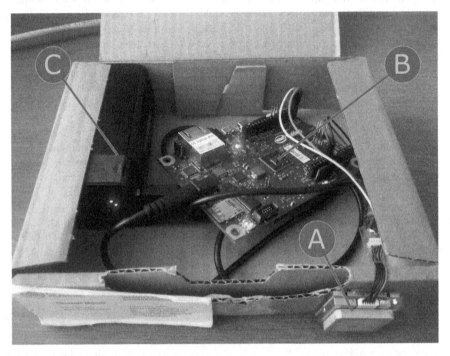

Fig. 7.5: Picture showing the smart object that was used to carry out the experiment. The smart object is composed by a GPS receiver (A), an Intel Galileo SBC running SMARTifact (B), and a powerbank (C).

To monitor this process, as shown in Figure 7.5, we equipped a cardboard box with an Intel Galileo SBC running SMARTifact and a GPS receiver, both powered by a powerbank, i.e., a rechargeable battery pack normally used to power and charge smartphones and tablets. Then, we put that box inside the briefcase the author takes with him when going to work, turning such a briefcase into a smart object. Once done so, the experiment was carried out as follows:

1. We modeled the VA-MI and MI-VA processes with BPMN process diagrams.
2. For one week, we relied on the smart briefcase to collect GPS data on the actual movement the author made when performing the processes.
3. By inspecting these GPS data, we identified a set of discrete states that the briefcase could assume, which were meaningful for the process. Using Node-

RED and the Geofence extension[12], we also defined the rules to derive the state of the briefcase from the GPS data, which are shown in Figure 7.6.

In particular, we extracted the position of the briefcase, together with the current date and time, from the NMEA strings [7] produced by the GPS receiver. Then, we discretized the position into eight states: [House], [VANord], and [VAFS] when the briefcase is in Varese near the author's house, Varese Nord and Varese FS railway stations, respectively. [MICadorna], [MIGaribaldi], [MIPiola], [MILambrate] and [Office] when the briefcase is in Milan near Milano Cadorna and Milano Garibaldi railway stations, Piola and Lambrate subway stations, and the author's office, respectively.

Also, we converted the date and time into a timestamp, and we embedded such a timestamp with the discrete states. Like in the previous experiment, we propagated the discretized states only if they were different from the previous ones.

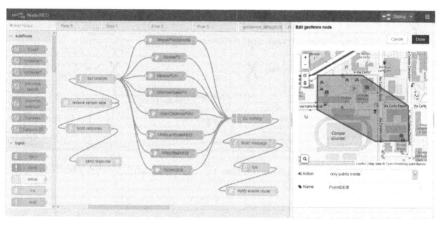

Fig. 7.6: Screenshot showing the rules to derive the state of the briefcase from the GPS sensor data modeled with Node-RED.

4. Following the method presented in Chapter 5, we enriched the BPMN process diagrams with artifacts representing the briefcase and its states. Figure 7.7 shows the enriched BPMN process diagram of the VA-MI (top) and MI-VA (bottom) processes. Then, we generated the E-GSM models and the artifact-to-object mapping criteria.

5. We instructed the SMARTifact platform running on the smart briefcase with the rules, process models and mapping criteria derived in the previous steps.

6. We used the smart briefcase to monitor the processes. In particular, the author powered the smart briefcase on every time he was going to work in his office, or he was going back home from there. The author also manually noted down the time when he started and finished each activity, if the process was not

---

[12] See https://www.npmjs.com/package/node-red-node-geofence.

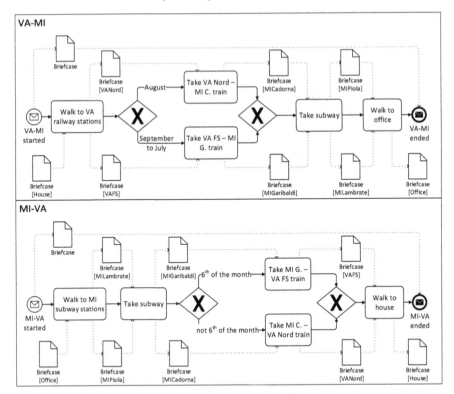

Fig. 7.7: BPMN process diagram of the VA-MI (top) and MI-VA (bottom) processes.

performed as expected (e.g., if, due to a strike or an accident, he had to take a different route) and, in such a case, which activities were affected.

7. Finally, we examined the results of the monitoring platform and compared them with the information noted down. In particular, to determine the completeness of each process execution, we verified that, whenever a stage changed its execution status, the author had actually started or finished an activity. Like in the previous experiment, we computed the detection delay for each activity activation and termination by computing the difference between the timestamp associated to the last discretized state causing a stage to change its status and the time when the author started or finished an activity. Finally, to identify incorrectly monitored executions, we checked if, for executions that were marked as correctly performed by the author, any *skipped*, *outOfOrder*, or *faulty* stage was present. Also, we checked if, for executions that were marked as incorrectly performed, *skipped*, *outOfOrder*, or *faulty* stages were present and, if so, were related to activities affected by the violations that occurred during execution.

Table 7.2: Results of the experiment on live sensor data.

Shipment	VA-MI	MI-VA	Global
Instances	18	19	37
Median duration [min]	96.76	107.33	102.05
Min. duration [min]	82.32	92.28	82.32
Max. duration [min]	116.55	122.37	122.37
Correctness [%]	94.44%	94.74%	94.59%
Completeness [%]	88.89%	94.74%	91.81%
Median detection delay [min]	−0.17	−0.95	−0.56
Median absolute d. delay [min]	1.72	2.11	1.91

The experiment lasted from May 10, 2017 to September 29, 2017, for a total of 37 process executions that lasted on average 102.05 minutes. Table 7.2 shows the results of this experiment. The monitoring platform was able to correctly determine the actual execution of a process for 94.59% of the total instances, which is in line with the results of the experiment on a simulated environment. In fact, only 2 executions, accounting for the 5.41% of the total, were incorrectly monitored. By analyzing the logs produced by the Onboard Sensor Gateway component, we found out that these issues were caused by the GPS receiver having difficulties in getting the signal. This caused the monitoring platform to be unable to detect when the author started walking to the railway stations or the subway stations.

Additionally, according to the notes taken by the author, every activity that was detected by the monitoring platform was actually executed. Also, the median detection delay amounted to −0.56 minutes (1.91 considering the absolute value of the delays). This is a sensible improvement with respect to the experiment on a simulated environment, which can be explained by two observations. Firstly, it was possible to tune the Onboard Sensor Gateway component to obtain sensor data adequate for reliably monitoring the process. Secondly, great diligence was observed by the author when noting down the activities being executed. On the other hand, when carrying out the experiment on a simulated environment, we had no control on the sensor data, and we had no information on how much diligent the truck drivers were in notifying when activities were executed.

7.4 Summary

This chapter demonstrated the applicability of the artifact-driven approach to monitor real processes. To do so, SMARTifact, an artifact-driven monitoring platform prototype, was developed according to the reference architecture presented in Section 3.3. SMARTifact was then tested against several processes belonging to the logistics domain. A first experiment focused on replaying historical sensor data provided by an European logistics company. A second experiment consisted in deploying SMARTifact on a real smart object, which was then used to monitor a live process. Further experiments, involving multiple smart objects communicating with each other, are currently being carried out in the context of the Italian project ITS Italy 2020.

The results of both experiments are quite encouraging. Over 90% of the total process executions were correctly monitored, and the incorrectly monitored ones were mainly caused by issues when collecting sensor data. Also, SMARTifact was capable of detecting when activities were executed with little delay with respect to manual notifications. The response time of SMARTifact, i.e. the time taken by the platform to process sensor data, was dependent on the throughput of the sensor data. Although not explicitly measured when the experiments were carried out, the response time was quite negligible when compared to the detection delay, as long as the GPS receiver was limited to one read per second.

Therefore, the results of these experiments show that the expectations that artifact-driven process monitoring brings are fulfilled: It is actually possible to autonomously detect when activities are carried out and when activities are not executed as expected. Also, when a violation occurs, the process continues to be monitored without human intervention.

Chapter 8
Conclusions

This book presented a novel technique, named artifact-driven process monitoring, to autonomously and continuously monitor the execution of business processes. This technique aimed at overcoming the limitation of traditional monitoring solutions, especially when dealing with human-centric and multi-party processes. In particular, with artifact-driven process monitoring, human operators are no longer required to send notifications to the monitoring platform whenever they start or conclude a business activity. Also, when a violation in the process execution occurs, human intervention is no longer required to continue monitoring the process.

Thank to the IoT paradigm, physical artifacts participating in a process were turned into smart objects. Being equipped with sensors, smart objects can be aware of their own state, represented by their physical conditions and the ones of the environment in which they operate. Based on this information, smart objects can autonomously identify when activities start or end, and if something wrong occurs when an activity is executed. Smart objects can also determine if they are misused, even if the process they participate in concludes successfully.

To provide guidelines on how to implement artifact-driven process monitoring, this book presented the reference architecture that a monitoring platform following this technique should use. In particular, four software components were defined: The On-board Sensor Gateway, to retrieve sensor data. The Events Processor, to determine the state of an artifact based on sensor data. The Events Router, to exchange information on the other artifacts. The E-GSM Engine, to keep track of the process to be monitored and to detect violations. This architecture fully exploited the IoT paradigm. Firstly, all these components can be deployed onto a smart object, making it capable to monitor a process alone. Secondly, the Events Router allowed smart objects to exchange information on their state with the other ones participating in the same process execution.

To be aware of the process to monitor, artifact-driven process monitoring requires a representation of the process. To this aim, this book presented an extension of the GSM artifact-centric language, named E-GSM. Imperative languages, such as BPMN, force the process to be executed exactly as specified in the model. As such, when used to monitor a process, imperative languages cause the monitoring platform to halt and wait for human intervention whenever a discrepancy between

G. Meroni: Artifact-Driven Business Process Monitoring, LNBIP 368, pp. 121–130, 2019
https://doi.org/10.1007/978-3-030-32412-4_8

the model and the actual execution occurs. E-GSM, on the other hand, treats dependencies among activities as descriptive rather than prescriptive. This way, whenever a discrepancy occurs, the monitoring platform simply flags the affected activities as incorrectly executed, without requiring any intervention.

Remarkably, GSM was extended by introducing two constructs: process flow guards and fault loggers. Process flow guards allow to model the dependencies among activities. This way, it is possible to detect activities that are executed when they should not, and activities that were not executed when they should. Fault loggers, on the other hand, define under which circumstance an activity fails to be executed. This way, it is possible to detect when something goes wrong while executing an activities.

To ease the configuration of an artifact-driven monitoring platform, this book introduced a method to derive from BPMN collaboration diagrams all the information required to drive the platform. The first step of this method consists in enriching collaboration diagrams with data objects representing the artifacts participating in the process, and the state they assume. This way, it is possible to define when smart objects, instantiating the artifacts when the process is executed, start and stop interacting with the process. Data objects also allow to specify the states that artifacts should have for activities to start or finish.

The second step consists in deriving, for each artifact participating in the process, a view on the process from the artifact's standpoint. To this aim, starting from the enriched collaboration diagram, a BPMN process diagram is derived for each artifact. In particular, activities that do not require the artifact to start or finish are omitted from the resulting process diagram.

The third and final step consists in deriving, from each BPMN process diagram obtained in the previous step, the information needed to configure the Events Router and E-GSM Engine components: *(i)* Criteria to dynamically bind and unbind the actual smart objects participating in a process execution to the corresponding artifacts. *(ii)* An E-GSM model representing the activities composing the process and their dependencies. *(iii)* An E-GSM model representing all the states that an artifact may assume when the process is executed, together with the expected transitions among these states.

To reliably monitor a process, artifact-driven process monitoring relies on the state of the artifacts participating in that process. As such, when enriching the process model with artifacts, designers should make sure that the smart objects can detect all the states indicated in the model. This book presented an ontology-based approach to automatically perform such an assessment.

Firstly, to quantify to which extent a process can be monitored, the monitorability metric was adopted. Then, the FIESTA-IoT and Physics Domain ontologies were adopted and extended to formalize the capabilities of the monitoring infrastructure, that is, the set of smart objects running the artifact-driven monitoring platform. Finally, formulas to compute the monitorability based on the information contained in the ontologies were introduced. Additionally, this book showed that, by opportunely querying the ontologies, suggestions can be provided to the designer on how to improve the monitorability.

To prove the applicability of artifact-driven process monitoring, a prototype of monitoring platform, named SMARTifact, was implemented according to the reference architecture. SMARTifact was then used to carry out two experiments on processes belonging to the logistics domain.

The first experiment relied on two datasets provided by a large European logistics company. The first dataset contained information on sensor data collected during the execution of shipping processes. The second dataset contained process logs produced by a traditional BPMS, based on manual notifications. These processes were simulated by feeding SMARTifact with sensor data. The monitoring results obtained by SMARTifact were then compared with the process logs.

The second experiment was carried out by deploying SMARTifact on a smart briefcase equipped with a GPS sensor. Such smart briefcase was then used to monitor the process the author of this book performs to reach his office from his house, and vice-versa. While the experiment was carried out, the author noted down when each activity was started or finished, and if the process was executed as expected. This allowed to compare the results produced by SMARTifact with such observations.

The results of both experiments showed that, in more than 90% of the total executions, SMARTifact was capable of correctly reporting violations. It was also observed that, on average, SMARTifact took only a few minutes to figure out when activities were started or completed, which is quite negligible for long-lasting processes. Finally, false positives, false negatives, or activities taking a long time to be detected were typically caused by issues on sensor data. As such, they could have been avoided by using different sensors. Therefore, artifact-driven process monitoring has proven to be a both feasible and quite reliable technique to monitor multi-party, human-centric business processes.

8.1 Answers to the Research Questions

In this section, we provide punctual answers to the research questions we presented in Chapter 1:

- **RQ1:** *How can process portions that are executed by different organizations be monitored?*
  Thank to the IoT paradigm, it is possible to deploy a process monitoring platform directly on the goods participating in the process and exchanged among the participants. This way, the monitoring platform can monitor the whole process, regardless of which organization performs which process portion.

  - **RQ1.1:** *How can each organization be aware of the execution of those process portions being carried out by the other organizations?*
    For a process to be performed, goods are required. Some of the goods are exchanged between these organizations. Additionally, to be executed, most activities require one or more goods to be present and physically accessible. Therefore, such goods can stay in close contact with the process, and can

cross the boundaries of the single organizations. By putting a monitoring platform on top of these goods, when goods are moved to another organization, the monitoring platform is moved as well. Thus, the owner of the goods can easily monitor process portions carried out outside its premises.

– **RQ1.2:** *Can a monitoring platform cross the boundaries of an organization?*
As long as the whole monitoring platform is confined inside the goods, it is no longer confined on the organization's premises. Thus, it can travel along with the goods, staying in close contact with the process.

– **RQ1.3:** *Can the IoT paradigm be used to decentralize process monitoring?*
The IoT paradigm makes possible to turn physical goods into smart objects. In particular, goods are equipped with sensors, a computing device and a communication interface. This way, they can collect information relevant for the process, process it, and communicate with the organizations. This makes possible to deploy a monitoring platform directly on the goods.

- **RQ2:** *How can organizations autonomously and continuously monitor business processes?*
Thank to a monitoring engine based on the E-GSM language, it is possible to autonomously and continuously monitor the process.

– **RQ2.1:** *How can incorrect executions of a process be detected as soon as they happen?*
The E-GSM provides two constructs, named Fault Logger and Process Flow Guard, to define under which conditions the process is not being performed as expected. Fault Loggers specify conditions or events that should never occur while a process portion is being executed. Thus, they allow to immediately detect which process portion is incorrectly executed. Process Flow Guards, on the other hand, indicate dependencies among activities that are expected to be fulfilled when the process is executed. Thus, whenever one of such dependencies is not fulfilled, the activities not respecting the dependencies are immediately detected.

– **RQ2.2:** *How can a process be monitored even after a violation occurred?*
Unlike imperative languages, like BPMN, E-GSM does not rely on dependencies among activities to detect when the next activity should be executed. Process Flow Guards are in fact descriptive constructs, not prescriptive ones. Thus, even though such dependencies are violated, an E-GSM-based monitoring engine is still capable of detecting when the activities are executed. Thus, it does not require any human intervention even after a violation occurs.

– **RQ2.3:** *Is it possible to determine the impact of a violation on the execution of the process?*
Thank to Process Flow Guards and Fault Loggers, an E-GSM-based monitoring engine can detect which process portions are affected by a violation. Additionally, it distinguishes between foreseen violations that were opportunely addressed, unforeseen violations, and foreseen violations that were ignored, from the least to the most severe. By opportunely assigning

weights to activities based on their importance, the effects of violations on the process can be then easily assessed.

- **RQ3:** *How can the conditions of the goods participating in a process be monitored?*
  By adopting the IoT paradigm, it is possible for the goods to autonomously monitor their own conditions. Thank to the E-GSM language, it is also possible to determine if the conditions of the goods change as expected.

  - **RQ3.1:** *Can the IoT paradigm allow physical goods to become self-aware of their conditions?*
    The IoT paradigm proved to be successful in making physical goods self-aware of their own conditions. In particular, thank to sensors, goods can determine their conditions. Thank to the computing device, they can reason on such conditions. Finally, thank to the communication interface, they can share this information with the organizations and the other goods.
  - **RQ3.2:** *Is it possible to determine if the goods are correctly manipulated? Can the expected evolution of the goods participating in a process be formalized? If so, is it possible to detect and promptly report violations in such an evolution?*
    E-GSM makes possible to model the evolution of the goods both in terms of activities being performed, and in terms of expected changes in the characteristics of the goods. This way, an E-GSM-based monitoring platform can detect if an activity was incorrectly performed, or if it was not executed at the right time. Also, the platform can detect if the characteristics of the goods change in an unexpected way. To ease such a formalization, it is also possible to semi-automatically derive the E-GSM models from the well-known BPMN collaboration diagrams. This way, process designers neither have to learn E-GSM, nor they have to model existing processes from scratch.
  - **RQ3.3:** *How can the identity of the goods participating in a process execution be specified?*
    Oftentimes, the identity of the goods participating in a process execution is known only after the process started. Therefore, it cannot be defined in advance in the process models. However, this information can be obtained by intercepting the messages exchanged between the information systems of the involved organizations. Those messages can be modeled as events in a BPMN model representing the process to monitor. Then, rules to capture these events and opportunely define when different smart objects start and stop interacting with the same process execution can be automatically extracted.
  - **RQ3.4:** *Which are the dependencies among the conditions of the goods and the execution of a process?*
    When a process is executed, the conditions of the goods participating in the process are expected to change in a precise and repeatable manner, as long as the process is performed as expected. Thus, by monitoring the conditions of the goods, it is possible to determine if the process is correctly

or incorrectly being executed. Additionally the conditions of the goods can be seen as requirements for activities to start, and as the outcome of the execution of activities.

- **RQ4:** *How can the execution of the activities be autonomously detected?*
  By adopting the E-GSM language to represent the process to monitor, the IoT paradigm to turn goods into smart objects, and ontologies to formalize the monitoring requirements, it is possible to autonomously and continuously monitor the process.

  - **RQ4.1:** *Is it possible to detect when the activities start or end without bothering operators?*
    The E-GSM language provides constructs, named Data Flow Guards and Milestones, that make possible to determine when activities are started or finished. In particular, Guards and Milestones can predicate on external data to determine when an activity is supposed to start or end. Thus, as long as these data can be obtained without bothering operators, it can be autonomously determined when activities are executed.
  - **RQ4.2:** *Can the goods manipulated during the execution of a process be used to determine when the activities are executed?*
    As long as the execution of an activity requires one or more goods to have certain characteristics, the activation of such an activity can be determined by the conditions of these goods. Similarly, if the execution of an activity changes the characteristics of one or more goods, the termination of such an activity can be determined by the conditions of these goods.
  - **RQ4.3:** *Can the IoT paradigm be used to detect when the activities start or end?*
    By exploiting the IoT paradigm, goods can become smart objects. Being equipped with sensors, such smart objects can become self-aware of their own conditions. Additionally, they can directly provide this information to an E-GSM engine, that can predicate on this information to detect when activities are executed.
  - **RQ4.4:** *Is it possible to assess to which extent the monitoring infrastructure is suited to monitor activities before the monitoring takes place? Is it possible to automatically provide suggestions on how to improve the monitoring infrastructure?*
    By using ontologies, it is possible to formalize the capabilities of the smart objects, that is, which physical characteristics can be sensed, and how can such characteristics be expressed. Once a formal model of the process to monitor has been defined, it is then possible to automatically determine which smart objects are suited to monitor that process. Also, suggestions on how to modify the other smart objects and/or the process model can be automatically provided.

8.2 Achievements in Runtime Process Monitoring

To compare the achievements of this book with the related work on conformance and compliance checking, the framework presented by Ly et al. in [73], which was introduced in Section 2.1.4, is adopted. Table 8.1 shows the functionalities supported by artifact-driven process monitoring, and how it positions with respect to some of the research work in this area.

Table 8.1: Comparison of the conformance and compliance monitoring functionalities (CMFs) supported by artifact-driven process monitoring, with respect to the related work (table adapted from [73]). The (+) mark indicates that the functionality is fully supported, (+/-) that it is partially supported, (-) that it is not supported, and (?) that it was not possible to determine.

Approach	CMF1	CMF2	CMF3	CMF4	CMF5	CMF6	CMF7	CMF8	CMF9	CMF10
Artifact-driven monitoring	+	+	+/-	+	+	+	+	-	+	+
Supervisory control theory [118]	+/-	-	+	+	+	-	-	+	-	-
ECE Rules [23]	+	+/-	+	+	-	-	+	-	+/-	+
BPath [120]	+	+	+	+	+/-	+	+	-	-	+/-
Dynamo [12, 13]	+	+	+/-	+	?	+	+	-	-	+/-
Gomez et al. [72]	+	-	-	+	?	+/-	+	+	-	-
Giblin et al. [44]	+	?	?	?	?	?	+	?	?	?
Narendra et al. [99]	-	+	+	?	-	+	+	-	-	+
Thullner et al. [125]	+	?	?	?	?	?	+	-	-	?
Mobucon LTL [78, 79]	+/-	-	-	+	-	-	+	+	+	+/-
MONPOLY [17]	+	+	+	+/-	+/-	+	+	-	-	-
Halle et al. [51]	+/-	+	+/-	?	?	?	+	?	?	?
Namiri et al. [98]	+/-	+	+	+	-	+	+	-	-	-
MobuconEC [92]	+	+	+	+	+	+	+	-	-	+/-
SeaFlows [74]	+/-	+/-	+/-	+	+	+	+	+	+	+/-

Artifact-driven process monitoring fully supports constraints on time (CMF1). Such constraints can be both qualitative, i.e., dependencies among activities, and quantitative, e.g. on the starting time or the duration of an activity. Constraints on data (CMF2) are also fully supported, while constraints on resources (CMF3) are supported only if the resources are implicitly modeled with data.

Since conditions on the activation, termination, and incorrect execution of activities are supported, artifact-driven process monitoring can predicate on the lifecycle of activities (CMF4). Also, it is possible to verify that activities were not executed as expected (CMF5), and to predicate on multiple instances of the same activity (CMF6).

Whenever a violation occurs, artifact-driven process monitoring can detect it almost immediately (CMF7). However, it is currently not possible to proactively detect violations (CMF8). As soon as a violation is detected, the affected portion of the process can be identified, and the component responsible for such a violation be determined (CMF9). Also, a metric to estimate how severely the process is affected by violations (CMF10) has been introduced.

8.3 Achievements in the Integration Among BPM and IoT

To frame the achievements of this book with respect to the synergies between the IoT and BPM, the challenges identified in the IoT-BPM manifesto [56] are adopted. By doing so, it is then possible to compare artifact-driven process monitoring with the related work in this area, that was identified in Section 2.3.3. Table 8.2 shows the results of such a comparison[1].

Table 8.2: Challenges on the integration of IoT with BPM that artifact-driven process monitoring addresses, compared to the related work (table inspired from [56]). The (+) mark indicates that the functionality is fully supported, (+/-) that it is partially supported, (-) that it is not supported.

Research work	IC1	IC2	IC4	IC5	IC6	IC7	IC12	IC13	IC14	IC16
Artifact-driven monitoring	+	+	+/-	+	+	+/-	+/-	+	+	+/-
Stertz et al. [122]	+/-	+	-	-	-	-	-	-	+/-	-
Mandal et al. [80]	-	+/-	-	+/-	+/-	+	+/-	+	+/-	-
Senderovich et al. [121]	-	+	-	-	-	-	+/-	+	+/-	+/-
Wombacher [133]	-	-	+	-	-	-	+/-	-	+	-
Weber et al. [130]	-	+/-	-	-	-	-	-	-	-	+
Gnimpieba et al. [45]	-	+	-	+	+/-	+/-	+/-	+	+/-	-
Knoch et al. [61]	+	+	+	-	+	-	-	+/-	+/-	+

To bridge the gap between sensor data and processes (IC13), the reference architecture that an artifact-driven monitoring platform should follow has been presented in Chapter 3. By adopting the IoT paradigm and the E-GSM language to represent the process to monitor, an artifact-driven process monitoring platform can detect when manual activities are executed without requiring humans to send notifications (IC2).

As discussed in Chapter 4, E-GSM is a declarative language and, as such, it allows to model semi-structured processes. This way, more flexibility that the one allowed by an imperative language, like BPMN, can be introduced in the processes to be monitored (IC5). In addition, E-GSM treats dependencies between activities

---

[1] Like in Section 2.3.3, challenges that are not related to business process monitoring are omitted.

as descriptive rather than prescriptive. This allows an artifact-driven monitoring platform to detect violations during execution as soon as they occur (IC14). Also, once a violation is detected, the platform can continue to monitor the process without requiring any intervention. Finally, thank to E-GSM, it is possible to determine under which conditions a task is incorrectly performed. This allows to predicate on the quality of the activities being executed (IC16). However, as the concept of resource has not been explicitly defined in this work, mechanisms to monitor their utilization are not available.

Thank to the method presented in Chapter 5, the process monitored by the smart objects is derived from the individual portions carried out by the participating organizations. This method also derives the expected lifecycle that each artifact should follow while the process is run, as well as rules to map smart objects to the artifacts participating in a specific process execution. This way, a linkage between these process portions, the expected evolution of the artifacts in the process, and the smart objects instantiating these artifacts is defined (IC6). Also, the complete end-to-end process model is broken into the portions carried out by the organizations, and the view that each artifact has on this process (IC7). However, the notions of habit and micro-process are not explicitly considered in this work.

Thank to the ontology-based approach presented in Chapter 6, suggestions on which sensors should be placed onto the smart objects instantiating the artifacts can be provided to the modeler (IC1). While not explicitly addressing process correctness check (IC4) and the detection of new situations (IC12), artifact-driven process monitoring can also be useful in this area. Indeed, the results obtained by a monitoring platform can provide insights on potential issues relate to the process model. For example, a huge amount of compliance violations on a specific process portion may indicate that this portion is incorrectly modeled. Similarly, reporting that an activity is never executed may indicate that the activity is no longer part of the process.

8.4 Current Limitations and Future Work

To autonomously detect when activities are executed, artifact-driven process monitoring expects these activities to interact with physical objects having certain characteristics. As such, when they interact only with non-tangible objects, such as invoices or purchase orders, manual notifications still have to be sent to know when such activities are executed. Additionally, to run an artifact-driven monitoring platform, physical objects have to be equipped with sensors, a SBC, and a communication interface. Depending on the type of these objects, this may not be economically or technically feasible. However, with the continuous improvements in the embedded technologies and the consequent price drop, in the future this issue may impact a lesser number of physical objects than today.

Future work will extend artifact-driven process monitoring with reactive mechanisms to mitigate violations. By equipping smart objects with actuators, it will be possible to automate the execution of activities. This way, whenever a violation is detected, smart objects will be able to directly take corrective actions.

Other aspects that will be tackled in the future are privacy and sensor data quality. In particular, policies to define to which extent organizations are entitled to know the execution of process portions will be introduced. A framework to specify the level of trust that can be put into each smart object, together with quality metrics (e.g., accuracy, veracity, etc.) on the data that sensors can provide, will be defined.

The ontology-based approach to quantify the monitorability of a process will be implemented in a tool. This tool, given a process model and the monitoring infrastructure, will assist the process designer in improving them. The ontologies will also be exploited to help organizations choose the smart objects that best monitor the process, and to automate the configuration of the monitoring infrastructure. Finally, artifact-driven process monitoring will be extended to processes relying on non-tangible objects.

Appendix A
Criteria to Evaluate the Integration Among BPM and IoT

To evaluate to which extent research work related to the integration among BPM and IoT addressed the challenges defined in [56], the following criteria were followed. As this book focuses on process monitoring, the evaluation criteria mainly focused on this aspect. Also, only integration challenges that could be related to process monitoring were considered.

A.1 Placing sensors in a process-oriented way (IC1)

Knowing how a business process is structured, and which data are needed by such a process can help to identify the optimum placement for sensors and smart objects.

We consider this challenge *fully addressed* only if, given a process model, it is possible to automatically determine which smart objects best support the process. We consider this challenge *partially addressed* if the process model is manually examined, and the choice of smart objects is not automated. If the process model is not taken into consideration at all when planning the IoT infrastructure, then this challenge is *not addressed*.

A.2 Monitoring manual activities (IC2)

Instead of requiring users to manually notify to a BPMS when activities are executed, data collected by smart objects can be used to automate such a task.

We consider this challenge *fully addressed* only if, thank to the smart objects, the execution of all activities composing a process can be automatically determined, and human intervention is no longer required for this task. We consider this challenge *partially addressed* if the execution of some activities is automatically determined, but human intervention is still required to fully monitor the process. If smart objects are not responsible for detecting when activities are executed, either directly or indirectly, then this challenge is *not addressed*.

A.3 Connecting analytical processes with the IoT (IC3)

To make business decisions, reliable and up to date information is required. Smart objects can help to integrate information traditionally available through databases and data warehouses with live data coming from sensors.

This challenge mainly focuses on supporting the execution of a process, rather than on monitoring how the process is performed. Therefore, this challenge was excluded from our assessment.

A.4 Exploiting the IoT to do process correctness check (IC4)

Sensor data collected by smart objects can help to identify issues in a process, such as deadlocks, livelocks or dead activities.

We consider this challenge *fully addressed* only if issues in the structure of the process are automatically derived from the data collected by the smart objects. We consider this challenge *partially addressed* if the data collected by smart objects can be used to detect structural issues, but they have to be manually inspected and interpreted. If the data collected by the smart objects is not taken into account when checking the structural correctness of a process, then this challenge is *not addressed*.

A.5 Dealing with unstructured environments (IC5)

Smart objects are often involved in ad-hoc processes, while BPM mostly deals with structured processes, where the structure and the interactions are at least partially known in advance. Therefore, new methodologies and techniques to deal with these ad-hoc processes are required.

We consider this challenge *fully addressed* only if a process-aware IoT-based solution supports completely ad-hoc processes, where dependencies among activities are completely optional. We consider this challenge *partially addressed* if the solution offers some degree of flexibility, but still requires the process to be partially structured. If only fully-structured processes are taken into account, then this challenge is *not addressed*.

A.6 Managing the links between micro processes (IC6)

Smart objects are oftentimes involved in micro processes representing habits, instead of complete end-to-end processes. Yet, these micro processes present links and dependencies among each other. Being able to detect and represent such dependencies is then required to have a complete overview.

We consider this challenge fully addressed only if multiple smart objects, each one in charge of monitoring or executing a process fragment, can autonomously cooperate to achieve a higher goal (e.g., monitoring the whole process). We consider this issue *partially addressed* if coordination aspects have to be managed before the solution is deployed. If no coordination among smart objects is foreseen, then this challenge is *not addressed*.

A.7 Breaking down end-to-end processes (IC7)

As mentioned before, smart objects participate in several micro processes that contribute to the execution of complete end-to-end processes. Being able to map portions of these end-to-end processes into micro processes is then required to properly execute them.

We consider this challenge *fully addressed* only if a process can be automatically broken into multiple fragments, which are then assigned to different smart objects. We consider this challenge *partially addressed* if the breakage of the process into fragments has to be manually performed. If the usage of process fragments is not taken into consideration, that is, only end-to-end processes are considered, then this challenge is *not addressed*.

A.8 Detecting new processes from data (IC8)

Sensor data provided by smart objects can be used to identify processes in a bottom-up fashion. As such, new processes can be discovered, and a certain degree of freedom in their execution can be introduced. However, constraints on such discovered processes should be introduced to optimize the usage of resources, and to prioritize specific goals.

This challenge is mainly related to process elicitation, that is, identifying the structure of a process that was not yet formalized. On the other hand, process monitoring expects the process model to be known. Therefore, this challenge was excluded from our assessment.

A.9 Specifying the autonomy level of smart objects (IC9)

In the IoT, smart objects autonomously react to events by executing tasks or process portions. While a certain level of autonomy is desirable, mechanisms to override such an autonomy in favor of a centralized supervision, or to veto certain actions should be introduced.

This challenge mainly focuses on automating and constraining the execution of a process, rather than on monitoring how the process is performed. Therefore, this challenge was excluded from our assessment.

A.10 Specifying the "social" roles of smart objects (IC10)

Since the goals of an organization's process may differ from the ones defined for the smart objects participating in such a process, governance mechanisms to resolve these conflicts are required.

This challenge mainly focuses on automating and constraining the execution of a process, rather than on monitoring how the process is performed. Therefore, this challenge was excluded from our assessment.

A.11 Concretizing abstract process models (IC11)

Oftentimes, business processes are firstly modeled in an abstract way, in order to capture the general behavior of the process, then dynamically turned into an executable model. To do so in the IoT context, matchmaking mechanisms that map activities or process portions of the abstract model to the smart objects according to their capabilities are required.

This challenge mainly focuses on automating and constraining the execution of a process, rather than on monitoring how the process is performed. Therefore, this challenge was excluded from our assessment.

A.12 Dealing with new situations (IC12)

To manage unplanned situations, BPM technologies such as task recommendation or the conditioned execution of processes and activities can be adopted. To this aim, sensor data collected by smart objects can provide more accurate information to contextualize such situations, and to compare them against previous ones.

We consider this challenge *fully addressed* only if mechanisms to identify new situations (e.g., by analyzing previous executions of a process) and opportunely change the process are explicitly supported. We consider this challenge *partially addressed* if, by manually analyzing information collected by an IoT-based platform, it is possible to identify new situations. If the IoT plays no role in the identification of new situations, then this challenge is *not addressed*.

A.13 Bridging the gap between process-based and event-based systems (IC13)

The IoT generates a large amount of heterogeneous sensor data. Being able to infer complex events from such data, and to correlate these events to process instances is far from trivial. To this aim, process mining techniques from one hand, and Complex Event Processor (CEP) from the other hand, can be helpful to bridge such a gap.

We consider this challenge *fully addressed* only if a clear separation between the logic to derive events relevant for a process, and the actual organization of the activities composing a process (i.e., the process model) exists. We consider this challenge *partially addressed* if some mechanism supporting the aggregation and correlation of events is made available. If the processing of sensor data is not considered at all, then this challenge is *not addressed*.

A.14 Improving online conformance checking (IC14)

As previously stated, sensor data collected by smart objects can be used to determine when activities are executed. This information can then be used to detect discrepancies between the process being executed and the planned one as soon as they occur.

We consider this challenge *fully addressed* only if, thank to the IoT, it is possible to continuously and autonomously detect violations in the execution order of the activities composing a process while the process is being executed. We consider this challenge *partially addressed* if human intervention is required either to identify the process portions affected by a violation, or to resume the monitoring after a violation occurs. If conformance cheeking occurs only after the process completes its execution, then this challenge is *not addressed*.

A.15 Improving resource utilization optimization (IC15)

In a pure IoT paradigm, the capabilities of smart objects are dimensioned according to the expected situations they have to react to. Exploiting BPM techniques, on the other hand, the capabilities can be optimized globally, with respect to the processes interacting with such smart objects.

This challenge mainly focuses on automating and constraining the execution of a process, rather than on monitoring how the process is performed. Therefore, this challenge was excluded from our assessment.

A.16 Improving resource monitoring and quality of task execution (IC16)

The IoT can help to identify issues in the utilization of resources participating in the process, such as excessive stress in human operators, or faults in machines.

We consider this challenge *fully addressed* only if, thank to the IoT, issues during the execution of activities can be detected. Also, suggestions on how to minimize the occurrence of such issues must be automatically provided to the user. We consider this challenge *partially addressed* if the quality of task execution can be qualitatively determined. If the quality of task execution cannot be determined, then this issue is *not addressed*.

Appendix B
BPMN to E-GSM Translation Proof of Correctness

B.1 Process Model

We provide a formal definition of (block-structured) BPMN process models manipulating artifact states. For simplicity, we consider only activities/tasks equipped with at most one boundary event.

B.1.1 Data Component

In our setting, the data component of a business process is constituted by a set of artifacts and their states. In particular, a data component is a set of pairs $\langle A, \Sigma \rangle$, where A is the name of an *artifact*, and Σ is the set of states in which that artifact can be. How an artifact moves from one state to another is implicitly determined by the process control-flow, and how atomic tasks operate over artifacts.

B.1.2 Blocks

Blocks account for the different units of work in the process, together with their control-flow dependencies. We therefore start by defining a generic notion of block, which then specializes depending on its type.

Definition B.1 (Block) A *block* is a triple $\langle BName, BType, BAttr \rangle$, where:
- *BName* is the *block name*, used to uniquely identify the block.
- $BType \in \{\texttt{event}, \texttt{proc}, \texttt{task}, \texttt{activity}, \texttt{seq}, \texttt{par}, \texttt{choice}, \texttt{or}, \texttt{loop}\}$ is the *block name*, where: *(i)* `event` denotes event blocks; *(ii)* `task` denotes task blocks; *(iii)* `activity` denotes non-atomic activities specified in terms of a subprocess; *(iv)* `seq` denotes sequence blocks (indicating the acceptable ordering of execution for other blocks); *(v)* `par` denotes parallel blocks (whose multiple branches are executed concurrently); *(vi)* `choice` denotes choice blocks (whose multiple

branches are mutually exclusive); *(vii)* or denotes inclusive or blocks (whose multiple branches are selectively executed in parallel); *(viii)* loop denotes loop blocks (where the flow may execute a block multiple times); *(ix)* proc denotes process blocks, which begin with a start event and finish with a termination event, provided that the process flow proceeds without exceptions in between.

- *BAttr* is a tuple of *type-dependent attributes*.

In the following, we detail how *BAttr* is defined depending on the block type. If *BAttr* contains another block B', we say that B' is a *direct sub-block* of B.

B.1.2.1 Event Block

An event block represents a BPMN start, intermediate, or termination event. In the context of this paper, the specific kind of event is not important, nor it is whether it is an event triggered by the process itself, or caught by the process. Hence, we simply consider $BAttr = \langle \rangle$.

B.1.2.2 Task Block

A task block represents an atomic unit of work within the process. In the context of this work, we keep track of how a task relates to the state of relevant artifacts. Specifically, $BAttr = \langle IS, OS \rangle$, where IS and SO are two sets respectively expressing the precondition and effect of the task in terms artifact states required by the task to start, and new states in which artifacts are moved when the task completes. Each entry in $IS \cup OS$, in turn, is a pair $\langle A, S \rangle$, where A is an artifact and S is a state.

B.1.2.3 Process Block

A process block represents a BPMN process, triggered by a start event, consisting of a main block, and finally ending with a termination event. Hence, $BAttr = \langle E_s, B, E_t \rangle$, where:

- E_s is an event block, accounting for the start event;
- B is a generic block, accounting for the main execution block of the process;
- E_t is an event block, accounting for the termination event.

B.1.2.4 Activity Block

An activity is a generic unit of work within the process. While it is executed, it may spawn interrupting or non-interrupting exceptional flows, in the case where it is equipped with boundary events catching external events, together with corresponding handlers. For simplicity, and without loss of generality, we consider here the case where each activity block is equipped with at most one boundary event.

In this light, we have $BAttr = \langle B, [EH] \rangle$, where:

- B is either a task or a process block, respectively representing the case where the activity is an atomic unit of work, or a non-atomic unit of work captured by a (sub)process.
- EH is an optional triple $\langle E, H, f \rangle$, where E is a (boundary) event block, H is a generic block modeling the handler of the event, and f is a flag indicating how the exception handler is managed. We have in particular three cases:
  - ($f = \texttt{nonint}$) the exception is managed in a *non-interrupting* way, i.e., H is run in parallel with the current block, synchronizing their termination;
  - ($f = \texttt{intfw}$) the exception is managed by *interrupting* the normal execution, then continuing the execution in a *forward* way;
  - ($f = \texttt{intbw}$) the exception is managed by *interrupting* the normal execution, then going *back* to re-execute the normal block again.

B.1.2.5 Sequence Block

Sequence blocks are made of 2 or more sub-blocks executed one after the other. We consequently have $BAttr = \langle B_1, \ldots, B_n \rangle$, where each B_i is a generic block, and $n \geq 2$. The ordering within $BAttr$ reflects the nature of the sequence. It is worth noting that a process block can be seen as a sequence block constituted by three sub-blocks, the first being the start event of the process, the second being the main execution block, and the third being the termination event block.

B.1.2.6 Parallel Blocks

Parallel blocks are constituted by 2 or more sub-blocks running concurrently. In this case, we then have that $BAttr = \langle B_1, \ldots, B_n \rangle$, where each B_i is a generic block, and $n \geq 2$.

B.1.2.7 Decision Blocks

Decision blocks are those in which one or more sub-blocks are executed depending on the result of some decision. In particular, each sub-block is *conditional*, in the sense that it is associated to a guard, and it is executed if the guard evaluates to true. A conditional block is a pair $\langle \varphi, B \rangle$, where φ is a condition, and B is a generic block. In the case of a decision block, $BAttr = \langle C_1, \ldots, C_n \rangle$, where $n \geq 2$, each $C_i = \langle \varphi_i, B_i \rangle$ is a conditional block, and the guards are such that $\bigvee_{i \in \{1, \ldots, n\}} \varphi_i = true$ (i.e., at least one condition evaluates to true). We have two types of decision block:

- `choice`, representing an exclusive choice block. In this case, we have that guards are pairwise disjoint, i.e., for every $i, j \in \{1, \ldots, n\}$ with $i \neq j$, $\varphi_i \wedge \varphi_j = false$. Combined with the assumption above, this means that exactly one guard evaluates to true.

- or, representing an inclusive or block. In this case, multiple guards may hold, leading to execute all the corresponding sub-blocks.

B.1.2.8 Loop Blocks

Loop blocks represent units of work that can be repeated multiple times. In this case, $BAttr = \langle FB, CB \rangle$, where FB is the block executed at least once when traversing the loop block, and $CB = \langle \varphi_{in}, B' \rangle$ is a conditional block executed when the loop condition φ_{in} evaluates to true. In this case, after executing CB the control-flow reiterates the execution of FB. If φ_{in} evaluates to false, then the loop is terminated.

B.1.3 Process Model

We are now in the position of defining what a process model is:

Definition B.2 (Process Model) A process model $\mathcal{B}$ is a pair $\langle D, P \rangle$, where D is a data component (cf. Section B.1.1), and P is a *process block* (defined in Section B.1.2.3) that represents the top-level, end-to-end BPMN process of interest. We assume that P obeys to the following assumptions:
- P is constituted by finitely many (direct and indirect) sub-blocks;
- no two sub-blocks of P have the same name.

The first requirement guarantees that the process model is finite. The second requirement simultaneously implies two properties: on the one hand, it makes the process model unambiguous; on the other hand, it makes the process model well-defined, in the sense that no block directly or indirectly embeds itself as a sub-block. Thanks to such assumptions, we get that the sub-block relation induces a tree-structure, rooted in the top-level, end-to-end process, and whose leaves are atomic tasks and events. More specifically, the sub-block relation forms a tree where P is the root, each other block has a unique parent block, and the leaves of the tree correspond to task or event blocks. We call this tree the *process tree* of $\mathcal{B}$ (or of P). Given a block B different than P, we use notation $B.Parent$ to identify its parent block. Furthermore, we respectively denote by $P.Blocks$, $P.Tasks$ and $P.Events$ the blocks present in the sub-tree rooted in P, the names of task blocks in $P.Tasks$, and the names of event blocks in $P.Tasks$.

The notion of parenthood is formally defined next.

Definition B.3 (Block Parenthood) Let $\langle D, P \rangle$ be a process model, and B_p, B_c two blocks in $P.Blocks$ different than P itself. We say that B_p *is the parent block of B_c, written $B_p = B_c.Parent$*, if one of the following conditions holds:
- $B_p = \langle \mathrm{n_p}, \mathtt{activity}, \langle B, EH \rangle \rangle$, with $B = B_c$, or with EH containing either a triple of the form $\langle B_c, H, f \rangle$ or of the form $\langle E, B_c, f \rangle$.
- $B_p = \langle \mathrm{n_p}, Type, \langle B_1, \ldots, B_n \rangle \rangle$, with $Type \in \{\mathtt{seq}, \mathtt{par}\}$ and some $B_i = B_c$.

- $B_p = \langle \mathtt{n_p}, \mathit{Type}, \langle CB_1, \ldots, CB_n \rangle \rangle$, with $\mathit{Type} \in \{\mathtt{choice}, \mathtt{or}\}$ and some $CB_i = \langle \varphi_i, B_c \rangle$.
- $B_p = \langle \mathtt{n_p}, \mathtt{loop}, \langle B, CB \rangle \rangle$, with either $B = B_c$, or $CB = \langle \varphi, B_c \rangle$.

Specularly, given a block B_p we define the set of children blocks of B_p, written $B_p.\mathit{Children}$, as the set $\{ B_c \mid B_c.\mathit{Parent} = B_p \}$.

B.2 Trace Conformance

In this section, we formally characterize when a trace over $\mathcal{B}$ *conforms to* $\mathcal{B}$, considering the control-flow semantics of the blocks contained in the process tree of $\mathcal{B}$.

Definition B.4 (Trace) A *trace* over a process model $\mathcal{B} = \langle D, P \rangle$ is a finite sequence $\langle \mathtt{t_1} \cdots \mathtt{t_n} \rangle$ of *steps*, where each step $\mathtt{t_i}$ has one of the following forms:
- (event execution) $\mathtt{t_i} = E$, where $E \in P.\mathit{Events}$.
- (task start) $\mathtt{t_i} = \langle \mathtt{start}, \mathtt{n} \rangle$, where $\mathtt{n}$ is the name of a task in $P.\mathit{Tasks}$.
- (task end) $\mathtt{t_i} = \langle \mathtt{end}, \mathtt{n} \rangle$, where $\mathtt{n}$ is the name of a task in $P.\mathit{Tasks}$.

In addition, $\mathtt{t_1} = E_s$, where E_s is the start event of P.

To define conformance, we first introduce a suitable notion of execution state, which in turn provides the basis for defining when an execution step is accepted, and what is the resulting execution state.

Definition B.5 (Execution state) An *execution state* over $\mathcal{B}$ is a pair $\langle \mathit{Curr}, \mathit{Next} \rangle$, where Curr is the set of *enacted* blocks in $\mathcal{B}$, and Next is the set of *enactable* blocks in $\mathcal{B}$.

Intuitively:
- Block B is *enacted* whenever there is an execution thread currently flowing through B.
- Block B is *enactable* if B becomes enacted in response to some execution step done in the current state of affairs.

Definition B.6 (Initial execution state) Let $\mathcal{B} = \langle D, P \rangle$ be a process model, with $P = \langle E, A, F \rangle$. The *initial execution state* of $\mathcal{B}$ is $\langle \emptyset, \{P, E\} \rangle$.

The initial execution state reflects the intuition that, at the beginning, no block is enacted, and the only enactable blocks are P itself together with its start event E.

Definition B.7 (Executable step) Given an execution state $s = \langle \mathit{Curr}, \mathit{Next} \rangle$ over process model $\mathcal{B}$, we say that step $\mathtt{t}$ *is executable in s* if one of the following conditions holds:
1. $\mathtt{t} = \mathtt{e}$ and $E \in \mathit{Next}$, where E is the event block named $\mathtt{e}$;
2. $\mathtt{t} = \langle \mathtt{start}, \mathtt{n} \rangle$ and $T \in \mathit{Next}$, where T is the task block named $\mathtt{n}$;
3. $\mathtt{t} = \langle \mathtt{end}, \mathtt{n} \rangle$ and $T \in \mathit{Curr}$, where T is the task block named $\mathtt{n}$.

Obviously, the actual execution of an executable step leads to update the state of its corresponding block. However, depending on where that block is located in the process model, this update could recursively trigger state updates for other blocks. We classify such updates in three categories:

- disablement of a block, resulting in the removal of the block from the set of enacted/enactable blocks;
- enactability of a block, resulting in the insertion of the block in the set of enactable blocks;
- enactment of a block, leading to move the block from the set of enactable blocks to that of enacted blocks.

Definition B.8 (Block disablement) Let $s = \langle Curr, Next \rangle$ be an execution state over $\mathcal{B}$, t a step executable in s, and B a block of $\mathcal{B}$. We say that B is *disabled by* t *in* s, or alternatively that t *disables* B *in* s, if one of the following conditions holds:

1. Task end execution step - base case:
   a. *(i)* $t = \langle \mathtt{end}, \mathtt{n} \rangle$, *(ii)* $B = \langle \mathtt{n}, \mathtt{task}, \langle IS, OS \rangle \rangle$, and *(iii)* $B \in Curr$. (A task end step disables its corresponding task block if such a block is currently in execution).

2. Event execution step - base cases:
   a. *(i)* $t = \mathtt{e}$, *(ii)* $B = \langle \mathtt{e}, \mathtt{event}, \langle \rangle \rangle$, and *(iii)* $B \in Next$. (An event step disables its corresponding event block if it is enactable).
   b. *(i)* $t = \mathtt{f}$, *(ii)* $B = \langle \mathtt{n}, \mathtt{proc}, \langle E_s, B', E_t \rangle \rangle$ with $E_t = \langle \mathtt{f}, \mathtt{event}, \langle \rangle \rangle$, and *(iii)* $E_t \in Next$. (An event step disables a process block if it corresponds to its termination event, and such termination event is enactable).
   c. *(i)* $t = \mathtt{e}$, *(ii)* B is a descendant of, or corresponds to, block B', where B' is the inner block of an activity block that in turn has an interrupting boundary event named $\mathtt{e}$, and *(iii)* $B' \in Curr$.
   (An event step disables all inner blocks of an enabled activity block having that event as interrupting exception).

3. Activity block - inductive cases:
   a. *(i)* $B = \langle \mathtt{n}, \mathtt{activity}, \langle B' \rangle \rangle$ *(ii)* B' is disabled by t. (An activity block without boundary events is disabled if its inner block is disabled by the given execution step).
   b. *(i)* $B = \langle \mathtt{n}, \mathtt{activity}, \langle B', \langle E, B'', \mathtt{nonint} \rangle \rangle \rangle$ *(ii)* B' is disabled by t; *(iii)* $B'' \notin Curr$. (An activity block with non-interrupting boundary event is disabled if its inner block is disabled and its event handler is not in execution).
   c. *(i)* $B = \langle \mathtt{n}, \mathtt{activity}, \langle B', \langle E, B'', \mathtt{nonint} \rangle \rangle \rangle$ *(ii)* B'' is disabled by t; *(iii)* $B' \notin Curr$. (An activity block with non-interupting boundary event is disabled if its event handler is disabled and its inner block is not in execution).
   d. *(i)* $B = \langle \mathtt{n}, \mathtt{activity}, \langle B', \langle E, B'', f \rangle \rangle \rangle$, with $f \in \{\mathtt{intfw}, \mathtt{intbw}\}$; *(ii)* B' is disabled by t. (An activity block with interupting boundary event is disabled if its inner block is disabled - this guarantees that its handler is not in execution).
   e. *(i)* $B = \langle \mathtt{n}, \mathtt{activity}, \langle B', \langle E, B'', f \rangle \rangle \rangle$, with $f \in \{\mathtt{intfw}, \mathtt{intbw}\}$; *(ii)* B'' is disabled by t. (An activity block with interupting boundary event is disabled

if its event handler is disabled - this guarantees that its inner block is not in execution).

4. Sequence block - inductive case: *(i)* $B = \langle \mathtt{n}, \mathtt{seq}, \langle B_1, \ldots, B_n \rangle \rangle$ *(ii)* B_n is disabled by $\mathtt{t}$. (A sequence block is disabled if its last inner block is disabled).

5. Parallel/or block - inductive case: *(i)* $B = \langle \mathtt{n}, type, \langle B_1, \ldots, B_n \rangle \rangle$, with $type \in \{\mathtt{par}, \mathtt{or}\}$ *(ii)* there exists $i \in \{1, \ldots, n\}$ such that B_i is disabled by $\mathtt{t}$ and for each $j \in \{1, \ldots, n\}$ with $j \neq i$, $B_j \notin Curr$. (A parallel/or block is disabled as soon as one of its inner block is disabled, and there is no other inner blocks currently enacted).

6. Choice block - inductive case: *(i)* $B = \langle \mathtt{n}, \mathtt{choice}, \langle B_1, \ldots, B_n \rangle \rangle$, with $type \in \{\mathtt{par}, \mathtt{or}\}$ *(ii)* there exists $i \in \{1, \ldots, n\}$ such that B_i is disabled by $\mathtt{t}$. (A choice block is disabled when one of its inner blocks is disabled - that block is the one that was selected when taking the choice).

7. Loop block - inductive case: *(i)* $B = \langle \mathtt{n}, \mathtt{loop}, \langle B_{fw}, \langle \varphi, B_{bw} \rangle \rangle \rangle$, *(ii)* B_{fw} is disabled by $\mathtt{t}$, *(iii)* φ is false in s. (A loop block is disabled when its forward inner block is disabled, and the loop condition evaluates to false).

We denote by $\mathcal{D}_s^{\mathtt{t}}$ the set of blocks in $\mathcal{B}$ that are disabled by $\mathtt{t}$ in s.

Definition B.9 (Block enactment) Let $s = \langle Curr, Next \rangle$ be an execution state over process model $\mathcal{B}$, $\mathtt{t}$ a step executable in s, and B a block of $\mathcal{B}$. We say that B *is enacted by* $\mathtt{t}$ *in* s, or alternatively that $\mathtt{t}$ *enacts* B *in* s, if $B \in Next$ and one of the following conditions holds:

1. Task execution step - base case *(i)* $\mathtt{t} = \langle \mathtt{start}, \mathtt{n} \rangle$; *(ii)* $B = \langle \mathtt{n}, \mathtt{task}, \langle \rangle \rangle$. (An enactable task is enacted when it gets started).

2. Event execution step - base case *(i)* $\mathtt{t} = \mathtt{e}$ *(ii)* B is the top process block, and its start event is $\mathtt{e}$. (The top process block is enacted when its start event occurs - this only applies to the top block, since for subprocesses the start event implicitly occurs whenever the flow reaches them).

3. Activity block - inductive case: *(i)* B is an activity block of the form $\langle \mathtt{n}, \mathtt{activity}, \langle B', [EH] \rangle \rangle$ *(ii)* B' is enacted by $\mathtt{t}$ in s. (An activity block is enacted when its inner task/process block is enacted).

4. Sequence block - inductive case: *(i)* B is a sequence block of the form $\langle \mathtt{n}, \mathtt{seq}, \langle B_1, \ldots, B_n \rangle \rangle$; *(ii)* B_1 is enacted by $\mathtt{t}$ in s. (A sequence block is enacted when its first inner block is enacted).

5. Gateway blocks - inductive case: *(i)* B is a gateway block of the form $\langle \mathtt{n}, type, \langle B_1, \ldots, B_n \rangle \rangle$ with $type \in \{\mathtt{par}, \mathtt{choice}, \mathtt{or}\}$; *(ii)* there exists $i \in \{1, \ldots, n\}$ such that B_i is enacted by $\mathtt{t}$ in s. (A gateway block is enacted as soon as one of its inner, enactable blocks is actually enacted - for decision blocks, inner blocks are enactable only if their guard condition is true, see below).

6. Choice blocks - inductive case: *(i)* B is a loop block of the form $\langle \mathtt{n}, \mathtt{loop}, \langle B', CB \rangle \rangle$; *(ii)* B' is enacted by $\mathtt{t}$ in s. (A loop block is enacted when its inner forward block is enacted).

We denote by $\mathcal{E}_s^{\mathtt{t}}$ the set of blocks in $\mathcal{B}$ that are enacted by $\mathtt{t}$ in s.

Definition B.10 (Block enactability)

Let $s = \langle Curr, Next \rangle$ be an execution state over process model $\mathcal{B}$, $\mathtt{t}$ a step executable in s, and B a block of $\mathcal{B}$. We say that B *is made enactable by* $\mathtt{t}$ *in* s, or alternatively that t *makes B enactable in s,* if $B \notin Next \cup Curr$ and one of the following conditions holds:

1. Event execution step - base cases:
   a. *(i)* $t = e$; *(ii)* the (top) process block of $\mathcal{B}$ is $P = \langle n, \mathtt{proc}, \langle E_s, B, E_t \rangle \rangle$, with $E_s = \langle e, \mathtt{event}, \langle \rangle \rangle$; *(iii)* $P \in Next$. (The occurrence of the start event of the top process block makes its inner block enactable).
   b. *(i)* $t = e$; *(ii)* there exists block $B_p = \langle n, \mathtt{activity}, \langle B', EH \rangle \rangle \in Curr$, where EH is of the form $\langle \langle e, \mathtt{event}, \langle \rangle \rangle, B, f \rangle$. (The occurence of the boundary event of an enacted activity makes the corresponding handler enactable).

2. Activity block - base case: *(i)* B is an event block; *(ii)* $B.Parent$ is an activity block having B as boundary event, i.e., it has the form $\langle n, \mathtt{activity}, \langle B_c, \langle B, H, f \rangle \rangle \rangle$ *(iii)* $B.Parent$ is enacted by t in s. (The enactment of an activity block makes its boundary event enactable).

3. Parent activity block - inductive case: *(i)* $B.Parent$ is an activity block having B as inner block, i.e., it has the form $\langle n, \mathtt{activity}, \langle B, \langle E, H, f \rangle \rangle \rangle$; *(ii)* $B.Parent$ is made enactable by t in s. (As soon as an activity block becomes enactable, its inner block becomes enactable as well).

4. Parent sequence block - inductive cases:
   a. *(i)* $B.Parent$ is a sequence block having B as first inner block, i.e., it has the form $\langle n, \mathtt{seq}, \langle B, \ldots \rangle \rangle$; *(ii)* $B.Parent$ is made enactable by t in s. (As soon as a sequence block becomes enactable, its first inner block becomes enactable as well).
   b. *(i)* $B.Parent$ is a sequence block having B as non-first inner block, i.e., it has the form $\langle n, \mathtt{seq}, \langle \ldots, B_p, B, \ldots \rangle \rangle$; *(ii)* $B.Parent \in Curr$; *(iii)* B_p is disabled by t in s. (The disablement of an inner block inside an enacted sequence makes the next inner block enactable).

5. Parent parallel block - inductive case: *(i)* $B.Parent$ is a parallel block having B as one of its inner blocks, i.e., it has the form $\langle n, \mathtt{par}, \langle \ldots, B, \ldots \rangle \rangle$; *(ii)* $B.Parent$ is made enactable by t in s. (As soon as a parallel block becomes enactable, its inner blocks becomes all enactable as well).

6. Parent decision block - inductive case: *(i)* $B.Parent$ is a decision block having B as one of its inner blocks, i.e., it has the form $\langle n, type, \langle \ldots, CB, \ldots \rangle \rangle$, with $type \in \{\mathtt{choice}, \mathtt{or}\}$ and $CB = \langle \varphi, B \rangle$; *(ii)* $B.Parent$ is made enactable by t in s; *(iii)* φ is true in s. (As soon as a decision block becomes enactable, its inner block(s) whose corresponding condition evaluates to true in the current state become(s) enactable as well).

7. Parent loop block - inductive cases:
   a. *(i)* $B.Parent$ is a loop block having B as its forward inner block, i.e., $B.Parent$ is of the form $\langle n, \mathtt{loop}, \langle B, CB \rangle \rangle$; *(ii)* $B.Parent$ is made enactable by t in s. (As soon as a loop block becomes enactable, its forward inner block becomes enactable as well).
   b. *(i)* $B.Parent$ is a loop block having B as its backward inner block, i.e., $B.Parent$ is of the form $\langle n, \mathtt{loop}, \langle FB, \langle \varphi, B \rangle \rangle \rangle$; *(ii)* the forward block FB of the loop is disabled by t in s; *(iii)* φ is true in s. (The backward inner block of a loop becomes enactable when the forward block of that loop gets

disabled, provided that the loop condition evaluates to true in the current state).

c. *(i)* $B.Parent$ is a loop block having B as its forward inner block, i.e., $B.Parent$ is of the form $\langle \mathtt{n}, \mathtt{loop}, \langle B, \langle \varphi, B' \rangle \rangle \rangle$; *(ii)* the backward block B' of the loop is disabled by $\mathtt{t}$ in s. (The forward inner block of a loop becomes enactable when the backward block of that loop gets disabled - this captures a new iteration of the loop).

We denote by $\mathcal{N}_s^\mathtt{t}$ the set of blocks in $\mathcal{B}$ that are made enactable by $\mathtt{t}$ in s.

With the three notions of block disablement, enactment and enactability at hand, we are now able to define the key notions of state update and, in turn, conformance.

Definition B.11 (State update)
  Let $s = \langle Curr, Next \rangle$ and $s' = \langle Curr', Next' \rangle$ be two execution states over process model $\mathcal{B}$, $\mathtt{t}$ a step executable in s, and B a block of $\mathcal{B}$. We say that $\mathtt{t}$ *updates s into s'*, or alternatively that s *is updated by* $\mathtt{t}$ *into s'*, if:
  1. $Curr' = (Curr \cup \mathcal{E}_s^\mathtt{t}) \setminus \mathcal{D}_s^\mathtt{t}$;
  2. $Next = (Next \cup \mathcal{N}_s^\mathtt{t}) \setminus \mathcal{D}_s^\mathtt{t}$;

Definition B.12 (Conformance) Given a process model $\mathcal{B}$, an execution trace $\tau = \langle \mathtt{t}_1, \ldots, \mathtt{t}_n \rangle$ over $\mathcal{B}$ *conforms to* $\mathcal{B}$ if there exists a sequence $\langle s_0, \ldots, s_n \rangle$ of execution states such that:
  1. s_0 is the initial execution state of $\mathcal{B}$;
  2. for every $i \in \{1, \ldots, n\}$, step $\mathtt{t}_i$ is executable in s_{i-1};
  3. for every $i \in \{1, \ldots, n\}$, step $\mathtt{t}_i$ updates s_{i-1} into s_i.

We remark that this definition of conformance faithfully reconstructs the classical notion of control-flow conformance defined by translating the process model of interest into a workflow net. In particular, enactability corresponds to the enablement of transitions, enactment to the firing of a transition, and state update to the calculation of the marking resulting from executing a transition in a previous marking.

B.3 Conformance Preservation of the Translation

We are now in the position of proving the main result of the paper. Recall that a process model $\mathcal{B}$ is transformed into a corresponding E-GSM model $\mathcal{G}_\mathcal{B}^P$ according to the transformation rules exhaustively defined in [14].[1] From now on, we formulate all our definitions and results using $\mathcal{B}$ to denote the BPMN process model of interest, and $\mathcal{G}_\mathcal{B}^P$ to refer to its corresponding E-GSM model.

As for E-GSM, we adopt the standard GSM execution semantics, with the extended lifecycle discussed in Section 4.2. This gives raise to the following definition of conformance in the E-GSM sense.

---

[1] See also the technical report available at https://re.public.polimi.it/handle/11311/976678.

Definition B.13 (E-GSM conformance) Given an E-GSM model $\mathcal{M}$, an execution trace τ over $\mathcal{M}$ *conforms to* $\mathcal{M}$ if the state resulting from the execution of τ over $\mathcal{M}$ is such that no stage of $\mathcal{M}$ is *out-of-order*.

We show that the transformation is correct, in the following precise sense.

Theorem B.1 *Given a process model $\mathcal{B}$ and an execution trace $\tau = \langle t_1, \ldots, t_n \rangle$ over $\mathcal{B}$, the following two conditions hold:*

1. *If τ conforms to $\mathcal{B}$, then τ conforms to $\mathcal{G}_\mathcal{B}^P$.*
2. *If τ does not conform to $\mathcal{B}$, let $\tau_p = \langle t_1, \ldots, t_k \rangle$ (with $k \leq n$) be the minimum prefix of τ such that τ_p does not conform to $\mathcal{B}$; then τ_p does not conform to $\mathcal{G}_\mathcal{B}^P$ either.*

Notice that the theorem expresses that τ conforms to $\mathcal{B}$ if and only if it conforms to $\mathcal{G}_\mathcal{B}^P$, with the additional property that in the case of non-conformance, then $\mathcal{G}_\mathcal{B}^P$ is able to detect a deviation as soon as it occurs.

To prove Theorem B.1, we show a stronger result. To formulate such a result, we first need to connect the execution state of a BPMN process model to that of the corresponding E-GSM model.

Definition B.14 (Corresponding state) Let $s = \langle Curr, Next \rangle$ be an execution state over $\mathcal{B}$, and σ be an execution state over $\mathcal{G}_\mathcal{B}^P$. We say that σ *corresponds to* s if the following conditions hold:

1. a process block B of $\mathcal{B}$ belongs to $Curr$ if and only if the corresponding stage S_B in $\mathcal{G}_\mathcal{B}^P$ is *regular* and *open* in σ;
2. a process block B of $\mathcal{B}$ belongs to $Next$ if and only if the corresponding stage S_B is *regular* and *closed*, and becomes *regular* and *open* upon the execution of a single execution step in σ.

With this notion at hand, we prove the following core result.

Lemma B.1 *Let s be an execution state over $\mathcal{B}$, σ be its corresponding state over $\mathcal{G}_\mathcal{B}^P$, and t be an execution step over $\mathcal{B}$. Then:*

1. *t is executable for $\mathcal{B}$ in s if and only if the execution of t in σ does not lead any stage of $\mathcal{G}_\mathcal{B}^P$ to become out-of-order.*
2. *if t is executable for $\mathcal{B}$ in s, then the actual execution of t over $\mathcal{B}$ in s leads to a state s' that corresponds to the state σ' obtained by executing t over $\mathcal{G}_\mathcal{B}^P$ in σ.*

It is straightforward to prove that Lemma B.1 implies Theorem B.1 (through its inductive application over the input trace τ).

To prove Lemma B.1, we proceed by induction on the execution state evolution, considering all the possible effects over the enacted and enactable blocks, depending on which blocks are affected by the current execution step. This, in turn, is captured by Definitions B.7, B.8, B.9, and B.10.

The base case is the initial state s_0. By definition, in this state there is no enacted block, and the only enactable blocks are the top process block together with its start event. Let $P = \langle n, \texttt{proc}, \langle E, A, F \rangle \rangle$ be the top process block of $\mathcal{B}$, and $E = \langle e, \texttt{event}, \langle \rangle \rangle$ be its start event block. Then $s_0 = \langle \emptyset, \langle P, E \rangle \rangle$. By construction, the initial state σ_0 of $\mathcal{G}_{\mathcal{B}}^P$ is such that there is no open stage, and the only stage that can be regularly opened is the main stage S_P corresponding to the main process block P, together with its first sub-stage S_E, corresponding to E. Consequently, our induction hypothesis applies, since σ_0 corresponds to s_0.

In this initial state, the only executable step for $\mathcal{B}$ is e, and the effect of its execution over s_0 is to obtain $s_1 = \langle \{P\}, \{A\} \rangle$. By considering the sequence block shown in Figure 5.7, this is also true for $\mathcal{G}_{\mathcal{B}}^P$. Assume that $t = e$. Then, t is executable in σ_0 according to $\mathcal{G}_{\mathcal{B}}^P$. In fact:

- t regularly opens stage S_P, because it makes true the data flow guard DFG1 of S_P, and no process-flow guard is present in S_P.
- t regularly opens and immediately closes the first inner stage S_E of S_P.

In addition, after executing t, the sub-stage S_A of S_P that corresponds to A is the only stage that can be open upon the execution of an additional step. In fact, it is the only sub-stage of S_P whose process-flow guard PFG1 is actually true (due to the fact that S_E has been closed). This shows that the execution state obtained by applying t over σ_0 for $\mathcal{G}_{\mathcal{B}}^P$ indeed corresponds to s_1.

Now assume that $t \neq e$, which is not executable in s_0 according to $\mathcal{B}$. We show that also $\mathcal{G}_{\mathcal{B}}^P$ forbids the execution of t. Since $t \neq e$, two cases may arise. Either t corresponds to the termination event of block F, or it corresponds to another task start/complete or event. We separately analyze the two cases.

If t corresponds to the termination event captured by F, then stage S_P regularly opens (since DFG3 becomes true), and S_F opens out-of-order, since its data-flow guard DFG1 is true, but its process flow guard PFG1 is false (due to the fact that stage S_A has not yet been opened, and hence has not yet achieved its milestone).

If t corresponds to the start/termination of a task, or to an event different from the start and termination events of P, t must trigger some sub-stage of S_A, In fact, S_A is the main stage of $\mathcal{G}_{\mathcal{B}}^P$, and hence directly or indirectly contains all events/tasks but the start and termination events of the top process. By the recursive definition of the transformation of $\mathcal{B}$ into $\mathcal{G}_{\mathcal{B}}^P$, this also means that the data-flow guard DFG1 of S_A becomes true. However, since the start event of P has not yet occurred, milestone M1 of stage S_E has not yet been achieved, and hence the process-flow guard PFG1 of S_A evaluates to false. This, in turn, means that S_A opens out-of-order.

This concludes the proof of the base case. As for the inductive case, let us consider an execution state $s = \langle Curr, Next \rangle$ over $\mathcal{B}$, the execution state σ over $\mathcal{G}_{\mathcal{B}}^P$ that corresponds to s, and an execution step t. We show that t is executable in s over $\mathcal{B}$ if and only if its execution in σ over $\mathcal{G}_{\mathcal{B}}^P$ does not cause any stage to become out-of-order. Then, we show that every executable step over $\mathcal{B}$ in s leads to a state s' that corresponds to the state σ' obtained after executing that step over $\mathcal{G}_{\mathcal{B}}^P$ in σ. To exhaustively prove this, we have to consider all the three possible types of execution steps, and all possible transitions a block may be subject to when moving from state s to s', which depend in turn on two factors: the type of the block, and the kind of transition the block is subject to. There are four possibilities for this latter dimension. Consider a block B. The four possibilites are: *(i)* the block is enactable in s and becomes enacted in s' (this is captured by Definition B.9); *(ii)* the block is enactable or enacted in s and is disabled in s' - hence, it is not enacted nor enactable in s' (this is captured by Definition B.8); *(iii)* the block is not enacted nor enactable in s, and it becomes enactable in s' (this is captured by Definition B.10); *(iv)* the block is not affected by the transition.

We discuss one such combinations, the others can be proven similarly. Consider as execution step the start of a task, i.e., $t = \langle \text{start}, a \rangle$, where a is the name of a task in $\mathcal{B}$, in turn identified as block $A = \langle a, \text{task}, \langle \rangle \rangle$ in $\mathcal{B}$. We use S_A to denote the stage of $\mathcal{G}_\mathcal{B}^P$ that correponds to A.

We prove that t is executable in s over $\mathcal{B}$ if and only if its execution in σ over $\mathcal{G}_\mathcal{B}^P$ does not lead any stage to become out-of-order.

($\Rightarrow$) According to Definition B.7, t is executable in s over $\mathcal{B}$ if and only if A is enactable in s. By Definition B.14, this implies that S_A can be regularly opened by means of a single execution step. By considering the transformation rule for tasks (c.f. Section 5.1.3.1), such an execution step is indeed t.

($\Leftarrow$) We prove the contrapositive, i.e., we show that if t is not executable in s over $\mathcal{B}$, then executing t in σ over $\mathcal{G}_\mathcal{B}^P$ causes at least one stage to open out-of-order. According to Definition B.7, t is not executable in s over $\mathcal{B}$ if and only if A is not enactable in s. By Definition B.14, this implies that S_A cannot be (regularly) opened by means of a single execution step. By inspecting the transformation rules in Section 5.1.3, one can immediately notice that the translation of every type of non-leaf block leads to a corresponding stage that reports, as data-flow guard, all the data-flow guards of its inner blocks. This, in turn, implies that there must be a stage S_P such that: *(i)* S_P has a data-flow guard that corresponds to the start of A, i.e., that matches with t; *(ii)* the parent stage of S_P is already open (in the extreme case, the parent is the top stage of $\mathcal{G}_\mathcal{B}^P$). By considering all possible structures for S_P, and the fact that S_A cannot be opened by t, one can show that there is always a sub-stage of S_P whose data-flow guard matches with t, and that has a process-flow guard evaluating to false. This, in turn, implies that, upon the execution of t, that stage will open out-of-order.

We now prove that, when t is executable in s, the actual execution of t in s and σ leads to consequent states s' and σ' that correspond to each other.

By inspecting Definition B.8, we notive that $\mathcal{D}_s^t = \emptyset$ (no block is disabled by the start of a task). Hence, we do not have to discuss the case of block disablement.

By inspecting Definition B.9, we have to discuss the following cases:

- (Definition B.9, Case 1) From the executability of t, we get that $A \in Next$, hence the task block A is enacted by t in s. Since σ corresponds to s, S_A can be open by a single execution step. Such step is actually t, since t matches the unique data-flow guard of S_A. Upon the execution of t, S_A becomes open in σ'. Hence, as far as A and S_A are concerned, σ' indeed corresponds to s'.

- An enactable ancestor block B in $\mathcal{B}$ (which actually corresponds to a stage S_B in $\mathcal{G}_\mathcal{B}^P$ containing the start of task a as one of its data-flow guards, hence opening when executing t) is enacted by t in s. We have to show, case-by-case, that for stage S_B all its process-flow guards evaluate to true, which in turn implies that S_B opens regularly.

  - (Definition B.9, Case 3) B is an activity block enacted by t. This case is trivial, since S_B in this case does not have any process-flow guard.

- (Definition B.9, Case 4) B is an enactable sequence block whose first inner block B' is enacted by t. Such a rule guarantees that S_B is opened regularly, since it does not have any process-flow guard. At the same time, it also guarantees that $S_{B'}$ is opened regularly. In fact, being $S_{B'}$ the first substage of S_B, its process-flow guards only checks that $S_{B'}$ is not opened twice while its parent stage S_B is open. This is indeed true in this case, since S_B and $S_{B'}$ are being opened simultaneously.
- (Definition B.9, Case 5) B is an enactable gateway block with one of its inner blocks B' enacted by t. By considering the rules in Section 5.1.3 that tackle the different gateway blocks, it is immediate to see that S_B is regularly enacted by t since it does not have any process-flow guard, whereas S'_B presents the same situation as the case discussed in the point above. The only subtle case is the one of choice. To show that this case is also properly handled, recall that, for a choice block, the fact that B' is enactable implies that the corresponding condition evaluated to true, and such a condition is also mentioned in the process-flow guard of S'_B.
- (Definition B.9, Case 6) B is an enactable loop block whose inner forward block B' is enacted by t. Again, also in this case S_B does not have any process-flow guard, and its substage $S_{B'}$ has a process-flow guard that simply checks that $S_{B'}$ has not yet been concluded.

By inspecting Definition B.10, the only case that must be discussed given the fact that t corresponds to the start of a task, is that of an activity block B that is enacted by t. In this situation, Case 2 of Definition B.10 applies, indicating that the boundary event of B becomes enactable. Indeed, they guarantee that upon the opening of S_B, special (fault logging) milestones corresponding to the boundary event are indeed achievable.

This concludes the proof for the case where the considered execution step is the start of a task. Task termination and event occurrence are proven similarly.

References

1. van der Aalst, W.M.P.: Business process management: A comprehensive survey. ISRN Software Engineering **2013**(507984), 37 (2013). doi: http://dx.doi.org/10.1155/2013/507984
2. van der Aalst, W.M.P., Pesic, M.: Decserflow: Towards a truly declarative service flow language. In: Bravetti, M., Núñez, M., Zavattaro, G. (eds.) Web Services and Formal Methods, Third International Workshop, WS-FM 2006 Vienna, Austria, September 8-9, 2006, Proceedings. Lecture Notes in Computer Science, vol. 4184, pp. 1–23. Springer (2006). doi: 10.1007/11841197_1, https://doi.org/10.1007/11841197_1
3. Adriansyah, A., Munoz-Gama, J., Carmona, J., van Dongen, B.F., van der Aalst, W.M.P.: Measuring precision of modeled behavior. Inf. Syst. E-Business Management **13**(1), 37–67 (2015). doi: 10.1007/s10257-014-0234-7, https://doi.org/10.1007/s10257-014-0234-7
4. Agarwal, R., Fernandez, D.G., Elsaleh, T., Gyrard, A., Lanza, J., Sánchez, L., Georgantas, N., Issarny, V.: Unified iot ontology to enable interoperability and federation of testbeds. In: 3rd IEEE World Forum on Internet of Things, WF-IoT 2016, Reston, VA, USA, December 12-14, 2016. pp. 70–75. IEEE Computer Society (2016). doi: 10.1109/WF-IoT.2016.7845470, https://doi.org/10.1109/WF-IoT.2016.7845470
5. Akyildiz, I.F., Su, W., Sankarasubramaniam, Y., Cayirci, E.: Wireless sensor networks: a survey. Computer networks **38**(4), 393–422 (2002)
6. Aloisio, G., Conte, D., Elefante, C., Epicoco, I., Marra, G.P., Mastrantonio, G., Quarta, G.: Sensorml for grid sensor networks. In: Arabnia, H.R. (ed.) Proceedings of the 2006 International Conference on Grid Computing & Applications, GCA 2006, Las Vegas, Nevada, USA, June 26-29, 2006. pp. 147–152. CSREA Press (2006)
7. Association, N.M.E., et al.: NMEA 0183–Standard for interfacing marine electronic devices. NMEA (2002)
8. Atzori, L., Iera, A., Morabito, G.: The internet of things: A survey. Computer networks **54**(15), 2787–2805 (2010)
9. Backmann, M., Baumgrass, A., Herzberg, N., Meyer, A., Weske, M.: Model-driven event query generation for business process monitoring. In: Lomuscio, A., Nepal, S., Patrizi, F., Benatallah, B., Brandic, I. (eds.) Service-Oriented Computing - ICSOC 2013 Workshops - CCSA, CSB, PASCEB, SWESE, WESOA, and PhD Symposium, Berlin, Germany, December 2-5, 2013. Revised Selected Papers. Lecture Notes in Computer Science, vol. 8377, pp. 406–418. Springer (2013). doi: 10.1007/978-3-319-06859-6_36, https://doi.org/10.1007/978-3-319-06859-6_36
10. Bandyopadhyay, S., Sengupta, M., Maiti, S., Dutta, S.: Role of middleware for internet of things: A study. International Journal of Computer Science and Engineering Survey **2**(3) (2011)
11. Baresi, L., Di Ciccio, C., Mendling, J., Meroni, G., Plebani, P.: martifact: an artifact-driven process monitoring platform. In: Clarisó et al. [30], http://ceur-ws.org/Vol-1920/BPM_2017_paper_188.pdf
12. Baresi, L., Guinea, S.: Dynamo: Dynamic monitoring of WS-BPEL processes. In: Benatallah, B., Casati, F., Traverso, P. (eds.) Service-Oriented Computing - ICSOC 2005, Third International Conference, Amsterdam, The Netherlands, December 12-15, 2005, Proceedings. Lecture Notes in Computer Science, vol. 3826, pp. 478–483. Springer (2005). doi: 10.1007/11596141_36, https://doi.org/10.1007/11596141_36
13. Baresi, L., Guinea, S., Pasquale, L.: Self-healing BPEL processes with dynamo and the jboss rule engine. In: Wolf, A.L. (ed.) Proceedings of the 2007 International Workshop on Engineering of Software Services for Pervasive Environments, ESSPE 2007, Dubrovnik, Croatia, September 4, 2007. pp. 11–20. ACM (2007). doi: 10.1145/1294904.1294906, http://doi.acm.org/10.1145/1294904.1294906
14. Baresi, L., Meroni, G., Plebani, P.: A gsm-based approach for monitoring cross-organization business processes using smart objects. In: Reichert and Reijers [111], pp. 389–400. doi: 10.1007/978-3-319-42887-1_32, https://doi.org/10.1007/978-3-319-42887-1_32
15. Baresi, L., Meroni, G., Plebani, P.: On handling business process anomalies through artifact-based modeling. In: España, S., Ivanovic, M., Savic, M. (eds.) Proceedings of the CAiSE'16

Forum, at the 28th International Conference on Advanced Information Systems Engineering (CAiSE 2016), Ljubljana, Slovenia, June 13-17, 2016. CEUR Workshop Proceedings, vol. 1612, pp. 9–16. CEUR-WS.org (2016), http://ceur-ws.org/Vol-1612/paper2.pdf

16. Baresi, L., Meroni, G., Plebani, P.: Using the guard-stage-milestone notation for monitoring bpmn-based processes. In: Schmidt, R., Guédria, W., Bider, I., Guerreiro, S. (eds.) Enterprise, Business-Process and Information Systems Modeling - 17th International Conference, BPMDS 2016, 21st International Conference, EMMSAD 2016, Held at CAiSE 2016, Ljubljana, Slovenia, June 13-14, 2016, Proceedings. Lecture Notes in Business Information Processing, vol. 248, pp. 18–33. Springer (2016). doi: 10.1007/978-3-319-39429-9_2, https://doi.org/10.1007/978-3-319-39429-9_2

17. Basin, D.A., Harvan, M., Klaedtke, F., Zalinescu, E.: MONPOLY: monitoring usage-control policies. In: Khurshid and Sen [60], pp. 360–364. doi: 10.1007/978-3-642-29860-8_27, https://doi.org/10.1007/978-3-642-29860-8_27

18. Basu, S., Pautasso, C., Zhang, L., Fu, X. (eds.): Service-Oriented Computing - 11th International Conference, ICSOC 2013, Berlin, Germany, December 2-5, 2013, Proceedings, Lecture Notes in Computer Science, vol. 8274. Springer (2013). doi: 10.1007/978-3-642-45005-1, https://doi.org/10.1007/978-3-642-45005-1

19. Baumgrass, A., Cabanillas, C., Di Ciccio, C.: A conceptual architecture for an event-based information aggregation engine in smart logitics. In: Kolb, J., Leopold, H., Mendling, J. (eds.) Enterprise modelling and information systems architectures. pp. 109–123. Gesellschaft für Informatik e.V., Bonn (2015)

20. Baumgrass, A., Di Ciccio, C., Dijkman, R.M., Hewelt, M., Mendling, J., Meyer, A., Pourmirza, S., Weske, M., Wong, T.Y.: GET controller and UNICORN: event-driven process execution and monitoring in logistics. In: Daniel, F., Zugal, S. (eds.) Proceedings of the BPM Demo Session 2015 Co-located with the 13th International Conference on Business Process Management (BPM 2015), Innsbruck, Austria, September 2, 2015. CEUR Workshop Proceedings, vol. 1418, pp. 75–79. CEUR-WS.org (2015), http://ceur-ws.org/Vol-1418/paper16.pdf

21. Baumgraß, A., Herzberg, N., Meyer, A., Weske, M.: BPMN extension for business process monitoring. In: Feltz, F., Mutschler, B., Otjacques, B. (eds.) Enterprise modelling and information systems architectures - EMISA 2014, Luxembourg, September 25-26, 2014. LNI, vol. 234, pp. 85–98. GI (2014), http://subs.emis.de/LNI/Proceedings/Proceedings234/article10.html

22. Beyer, J., Kuhn, P., Hewelt, M., Mandal, S., Weske, M.: Unicorn meets chimera: Integrating external events into case management. In: Azevedo, L., Cabanillas, C. (eds.) Proceedings of the BPM Demo Track 2016 Co-located with the 14th International Conference on Business Process Management (BPM 2016), Rio de Janeiro, Brazil, September 21, 2016. CEUR Workshop Proceedings, vol. 1789, pp. 67–72. CEUR-WS.org (2016), http://ceur-ws.org/Vol-1789/bpm-demo-2016-paper13.pdf

23. Bragaglia, S., Chesani, F., Mello, P., Montali, M., Sottara, D.: Fuzzy conformance checking of observed behaviour with expectations. In: Pirrone, R., Sorbello, F. (eds.) AI*IA 2011: Artificial Intelligence Around Man and Beyond - XIIth International Conference of the Italian Association for Artificial Intelligence, Palermo, Italy, September 15-17, 2011. Proceedings. Lecture Notes in Computer Science, vol. 6934, pp. 80–91. Springer (2011). doi: 10.1007/978-3-642-23954-0_10, https://doi.org/10.1007/978-3-642-23954-0_10

24. Cabanillas, C., Di Ciccio, C̄., Mendling, J., Baumgrass, A.: Predictive task monitoring for business processes. In: Sadiq, S.W., Soffer, P., Völzer, H. (eds.) Business Process Management - 12th International Conference, BPM 2014, Haifa, Israel, September 7-11, 2014. Proceedings. Lecture Notes in Computer Science, vol. 8659, pp. 424–432. Springer (2014). doi: 10.1007/978-3-319-10172-9_31, https://doi.org/10.1007/978-3-319-10172-9_31

25. Cai, H., Xu, L.D., Xu, B., Xie, C., Qin, S., Jiang, L.: Iot-based configurable information service platform for product lifecycle management. IEEE Trans. Industrial Informatics 10(2), 1558–1567 (2014). doi: 10.1109/TII.2014.2306391, https://doi.org/10.1109/TII.2014.2306391

26. Cassar, G., Barnaghi, P.M., Wang, W., Moessner, K.: A hybrid semantic matchmaker for iot services. In: 2012 IEEE International Conference on Green Computing and Communications, Conference on Internet of Things, and Conference on Cyber, Physical and

Social Computing, GreenCom/iThings/CPSCom 2012, Besancon, France, November 20-23, 2012. pp. 210–216. IEEE Computer Society (2012). doi: 10.1109/GreenCom.2012.40, https://doi.org/10.1109/GreenCom.2012.40

27. Chabridon, S., Laborde, R., Desprats, T., Oglaza, A., Marie, P., Marquez, S.M.: A survey on addressing privacy together with quality of context for context management in the internet of things. Annales des Télécommunications **69**(1-2), 47–62 (2014). doi: 10.1007/s12243-013-0387-2, https://doi.org/10.1007/s12243-013-0387-2

28. Chesani, F., Lamma, E., Mello, P., Montali, M., Riguzzi, F., Storari, S.: Exploiting inductive logic programming techniques for declarative process mining. Trans. Petri Nets and Other Models of Concurrency **2**, 278–295 (2009). doi: 10.1007/978-3-642-00899-3_16, https://doi.org/10.1007/978-3-642-00899-3_16

29. Chinosi, M., Trombetta, A.: BPMN: an introduction to the standard. Computer Standards & Interfaces **34**(1), 124–134 (2012). doi: 10.1016/j.csi.2011.06.002, https://doi.org/10.1016/j.csi.2011.06.002

30. Clarisó, R., Leopold, H., Mendling, J., van der Aalst, W.M.P., Kumar, A., Pentland, B.T., Weske, M. (eds.): Proceedings of the BPM Demo Track and BPM Dissertation Award co-located with 15th International Conference on Business Process Modeling (BPM 2017), Barcelona, Spain, September 13, 2017, CEUR Workshop Proceedings, vol. 1920. CEUR-WS.org (2017), http://ceur-ws.org/Vol-1920

31. Coccoli, M., Torre, I.: Interacting with annotated objects in a semantic web of things application. Journal of Visual Languages & Computing **25**(6), 1012–1020 (2014)

32. Compton, M., Barnaghi, P.M., Bermudez, L., Garcia-Castro, R., Corcho, Ó., Cox, S.J.D., Graybeal, J., Hauswirth, M., Henson, C.A., Herzog, A., Huang, V.A., Janowicz, K., Kelsey, W.D., Phuoc, D.L., Lefort, L., Leggieri, M., Neuhaus, H., Nikolov, A., Page, K.R., Passant, A., Sheth, A.P., Taylor, K.: The SSN ontology of the W3C semantic sensor network incubator group. J. Web Sem. **17**, 25–32 (2012). doi: 10.1016/j.websem.2012.05.003, https://doi.org/10.1016/j.websem.2012.05.003

33. Corredor, I., Martínez, J.F., Familiar, M.S., López, L.: Knowledge-aware and service-oriented middleware for deploying pervasive services. Journal of Network and Computer Applications **35**(2), 562–576 (2012)

34. Dahanayake, A., Welke, R.J., Cavalheiro, G.: Improving the understanding of BAM technology for real-time decision support. IJBIS **7**(1), 1–26 (2011). doi: 10.1504/IJBIS.2011.037294, https://doi.org/10.1504/IJBIS.2011.037294

35. Damaggio, E., Hull, R., Vaculín, R.: On the equivalence of incremental and fixpoint semantics for business artifacts with guard-stage-milestone lifecycles. Information Systems **38**(4), 561–584 (2013). doi: 10.1016/j.is.2012.09.002, https://doi.org/10.1016/j.is.2012.09.002

36. Di Ciccio, C., van der Aa, H., Cabanillas, C., Mendling, J., Prescher, J.: Detecting flight trajectory anomalies and predicting diversions in freight transportation. Decision Support Systems **88**, 1–17 (2016). doi: 10.1016/j.dss.2016.05.004, https://doi.org/10.1016/j.dss.2016.05.004

37. Di Ciccio, C., Maggi, F.M., Montali, M., Mendling, J.: Ensuring model consistency in declarative process discovery. In: Motahari-Nezhad et al. [93], pp. 144–159. doi: 10.1007/978-3-319-23063-4_9, https://doi.org/10.1007/978-3-319-23063-4_9

38. Dumas, M., La Rosa, M., Mendling, J., Reijers, H.A.: Fundamentals of Business Process Management. Springer (2013). doi: 10.1007/978-3-642-33143-5, https://doi.org/10.1007/978-3-642-33143-5

39. Eid-Sabbagh, R., Hewelt, M., Meyer, A., Weske, M.: Deriving business process data architectures from process model collections. In: Basu et al. [18], pp. 533–540. doi: 10.1007/978-3-642-45005-1_43, https://doi.org/10.1007/978-3-642-45005-1_43

40. Engel, R., van der Aalst, W.M.P., Zapletal, M., Pichler, C., Werthner, H.: Mining inter-organizational business process models from EDI messages: A case study from the automotive sector. In: Ralyté et al. [110], pp. 222–237. doi: 10.1007/978-3-642-31095-9_15, https://doi.org/10.1007/978-3-642-31095-9_15

41. Eshuis, R., Gorp, P.V.: Synthesizing data-centric models from business process models. Computing **98**(4), 345–373 (2016). doi: 10.1007/s00607-015-0442-0, https://doi.org/10.1007/s00607-015-0442-0

42. Fahland, D., Lübke, D., Mendling, J., Reijers, H.A., Weber, B., Weidlich, M., Zugal, S.: Declarative versus imperative process modeling languages: The issue of understandability. In: Halpin, T.A., Krogstie, J., Nurcan, S., Proper, E., Schmidt, R., Soffer, P., Ukor, R. (eds.) Enterprise, Business-Process and Information Systems Modeling, 10th International Workshop, BPMDS 2009, and 14th International Conference, EMMSAD 2009, held at CAiSE 2009, Amsterdam, The Netherlands, June 8-9, 2009. Proceedings. Lecture Notes in Business Information Processing, vol. 29, pp. 353–366. Springer (2009). doi: 10.1007/978-3-642-01862-6_29, https://doi.org/10.1007/978-3-642-01862-6_29

43. Francescomarino, C.D., Marchetto, A., Tonella, P.: Reverse engineering of business processes exposed as web applications. In: Winter, A., Ferenc, R., Knodel, J. (eds.) 13th European Conference on Software Maintenance and Reengineering, CSMR 2009, Architecture-Centric Maintenance of Large-SCale Software Systems, Kaiserslautern, Germany, 24-27 March 2009. pp. 139–148. IEEE Computer Society (2009). doi: 10.1109/CSMR.2009.26, https://doi.org/10.1109/CSMR.2009.26

44. Giblin, C.J., Mueller, S., Pfitzmann, B.: From regulatory policies to event monitoring rules: towards model-driven compliance automation. Tech. rep., IBM (2006)

45. Gnimpieba, Z.D.R., Nait-Sidi-Moh, A., Durand, D., Fortin, J.: Using internet of things technologies for a collaborative supply chain: Application to tracking of pallets and containers. In: The 10th International Conference on Future Networks and Communications (FNC 2015) / The 12th International Conference on Mobile Systems and Pervasive Computing (MobiSPC 2015) / Affiliated Workshops, August 17-20, 2015, Belfort, France. Procedia Computer Science, vol. 56, pp. 550–557. Elsevier (2015). doi: 10.1016/j.procs.2015.07.251, https://doi.org/10.1016/j.procs.2015.07.251

46. Guarino, N., Carrara, M., Giaretta, P.: An ontology of meta-level categories. In: Doyle, J., Sandewall, E., Torasso, P. (eds.) Proceedings of the 4th International Conference on Principles of Knowledge Representation and Reasoning (KR'94). Bonn, Germany, May 24-27, 1994. pp. 270–280. Morgan Kaufmann (1994)

47. Güceglioglu, A.S., Demirörs, O.: A process based model for measuring process quality attributes. In: Richardson, I., Abrahamsson, P., Messnarz, R. (eds.) Software Process Improvement, 12th European Conference, EuroSPI 2005, Budapest, Hungary, November 9-11, 2005, Proceedings. Lecture Notes in Computer Science, vol. 3792, pp. 118–129. Springer (2005). doi: 10.1007/11586012_12, https://doi.org/10.1007/11586012_12

48. Guinard, D., Trifa, V., Karnouskos, S., Spiess, P., Savio, D.: Interacting with the soa-based internet of things: Discovery, query, selection, and on-demand provisioning of web services. IEEE Transactions on Services Computing 3(3), 223–235 (Jul 2010). doi: 10.1109/TSC.2010. 3

49. Gyrard, A., Datta, S.K., Bonnet, C., Boudaoud, K.: Cross-domain internet of things application development: M3 framework and evaluation. In: Awan, I., Younas, M., Mecella, M. (eds.) 3rd International Conference on Future Internet of Things and Cloud, FiCloud 2015, Rome, Italy, August 24-26, 2015. pp. 9–16. IEEE Computer Society (2015). doi: 10.1109/FiCloud.2015.10, https://doi.org/10.1109/FiCloud.2015.10

50. Hachem, S., Teixeira, T., Issarny, V.: Ontologies for the internet of things. In: Proceedings of the 8th Middleware Doctoral Symposium. pp. 3:1–3:6. MDS '11, ACM, New York, NY, USA (2011). doi: 10.1145/2093190.2093193, http://doi.acm.org/10.1145/2093190.2093193

51. Hallé, S., Villemaire, R.: Runtime monitoring of message-based workflows with data. In: 12th International IEEE Enterprise Distributed Object Computing Conference, ECOC 2008, 15-19 September 2008, Munich, Germany. pp. 63–72. IEEE Computer Society (2008). doi: 10.1109/EDOC.2008.32, https://doi.org/10.1109/EDOC.2008.32

52. Herzberg, N., Meyer, A., Weske, M.: Improving business process intelligence by observing object state transitions. Data Knowl. Eng. 98, 144–164 (2015). doi: 10.1016/j.datak.2015. 07.008, https://doi.org/10.1016/j.datak.2015.07.008

53. Hull, R., Damaggio, E., Fournier, F., Gupta, M., III, F.F.T.H., Hobson, S., Linehan, M.H., Maradugu, S., Nigam, A., Sukaviriya, P., Vaculín, R.: Introducing the guard-stage-milestone approach for specifying business entity lifecycles. In: Bravetti, M., Bultan, T. (eds.) Web Services and Formal Methods - 7th International Workshop, WS-FM 2010, Hoboken, NJ, USA, September 16-17, 2010. Revised Selected Papers. Lecture Notes in Computer Science,

vol. 6551, pp. 1–24. Springer (2010). doi: 10.1007/978-3-642-19589-1_1, https://doi.org/10.1007/978-3-642-19589-1_1

54. III, F.F.T.H., Boaz, D., Gupta, M., Vaculín, R., Sun, Y., Hull, R., Limonad, L.: Barcelona: A design and runtime environment for declarative artifact-centric BPM. In: Basu et al. [18], pp. 705–709. doi: 10.1007/978-3-642-45005-1_65, https://doi.org/10.1007/978-3-642-45005-1_65

55. Issarny, V., Georgantas, N., Hachem, S., Zarras, A., Vassiliadist, P., Autili, M., Gerosa, M.A., Hamida, A.B.: Service-oriented middleware for the future internet: state of the art and research directions. Journal of Internet Services and Applications 2(1), 23–45 (2011)

56. Janiesch, C., Koschmider, A., Mecella, M., Weber, B., Burattin, A., Di Ciccio, C., Gal, A., Kannengiesser, U., Mannhardt, F., Mendling, J., Oberweis, A., Reichert, M., Rinderle-Ma, S., Song, W., Su, J., Torres, V., Weidlich, M., Weske, M., Zhang, L.: The internet-of-things meets business process management: Mutual benefits and challenges (2017), paper available at https://arxiv.org/abs/1709.03628

57. Jorbina, K., Rozumnyi, A., Verenich, I., Francescomarino, C.D., Dumas, M., Ghidini, C., Maggi, F.M., Rosa, M.L., Raboczi, S.: Nirdizati: A web-based tool for predictive process monitoring. In: Clarisó et al. [30], http://ceur-ws.org/Vol-1920/BPM_2017_paper_202.pdf

58. Jouault, F., Allilaire, F., Bézivin, J., Kurtev, I.: ATL: A model transformation tool. Sci. Comput. Program. 72(1-2), 31–39 (2008). doi: 10.1016/j.scico.2007.08.002, https://doi.org/10.1016/j.scico.2007.08.002

59. Kang, B., Kim, D., Kang, S.: Real-time business process monitoring method for prediction of abnormal termination using knni-based LOF prediction. Expert Syst. Appl. 39(5), 6061–6068 (2012). doi: 10.1016/j.eswa.2011.12.007, https://doi.org/10.1016/j.eswa.2011.12.007

60. Khurshid, S., Sen, K. (eds.): Runtime Verification - Second International Conference, RV 2011, San Francisco, CA, USA, September 27-30, 2011, Revised Selected Papers, Lecture Notes in Computer Science, vol. 7186. Springer (2012). doi: 10.1007/978-3-642-29860-8, https://doi.org/10.1007/978-3-642-29860-8

61. Knoch, S., Ponpathirkoottam, S., Fettke, P., , Loos, P.: Technology-enhanced process elicitation of worker activities in manufacturing (2017), to appear on Business Process Management Workshops - BPM 2017 International Workshops

62. Köpke, J., Su, J.: Towards ontology guided translation of activity-centric processes to GSM. In: Reichert and Reijers [111], pp. 364–375. doi: 10.1007/978-3-319-42887-1_30, https://doi.org/10.1007/978-3-319-42887-1_30

63. Köpke, J., Su, J.: Towards quality-aware translations of activity-centric processes to guard stage milestone. In: La Rosa, M., Loos, P., Pastor, O. (eds.) Business Process Management - 14th International Conference, BPM 2016, Rio de Janeiro, Brazil, September 18-22, 2016. Proceedings. Lecture Notes in Computer Science, vol. 9850, pp. 308–325. Springer (2016). doi: 10.1007/978-3-319-45348-4_18, https://doi.org/10.1007/978-3-319-45348-4_18

64. Kotis, K., Katasonov, A.: Semantic interoperability on the web of things: The semantic smart gateway framework. In: Barolli, L., Xhafa, F., Vitabile, S., Uehara, M. (eds.) Sixth International Conference on Complex, Intelligent, and Software Intensive Systems, CISIS 2012, Palermo, Italy, July 4-6, 2012. pp. 630–635. IEEE Computer Society (2012). doi: 10.1109/CISIS.2012.200, https://doi.org/10.1109/CISIS.2012.200

65. Krakowiak, S.: What's middleware? ObjectWeb. org (2003)

66. Kumaran, S., Liu, R., Wu, F.Y.: On the duality of information-centric and activity-centric models of business processes. In: Bellahsene, Z., Léonard, M. (eds.) Advanced Information Systems Engineering, 20th International Conference, CAiSE 2008, Montpellier, France, June 16-20, 2008, Proceedings. Lecture Notes in Computer Science, vol. 5074, pp. 32–47. Springer (2008). doi: 10.1007/978-3-540-69534-9_3, https://doi.org/10.1007/978-3-540-69534-9_3

67. Kurz, M., Schmidt, W., Fleischmann, A., Lederer, M.: Leveraging CMMN for ACM: examining the applicability of a new OMG standard for adaptive case management. In: Ehlers, J., Thalheim, B. (eds.) Proceedings of the 7th International Conference on Subject-Oriented Business Process Management, S-BPM ONE 2015, Kiel, Germany, April 23-24, 2015. pp. 4:1–4:9. ACM (2015). doi: 10.1145/2723839.2723843, http://doi.acm.org/10.1145/2723839.2723843

68. La Rosa, M., Soffer, P. (eds.): Business Process Management Workshops - BPM 2012 International Workshops, Tallinn, Estonia, September 3, 2012. Revised Papers, Lecture Notes in Business Information Processing, vol. 132. Springer (2013). doi: 10.1007/978-3-642-36285-9, https://doi.org/10.1007/978-3-642-36285-9
69. Le-Phuoc, D., Nguyen-Mau, H.Q., Parreira, J.X., Hauswirth, M.: A middleware framework for scalable management of linked streams. Web Semantics: Science, Services and Agents on the World Wide Web **16**, 42–51 (2012)
70. de Leoni, M., van der Aalst, W.M.P.: Aligning event logs and process models for multi-perspective conformance checking: An approach based on integer linear programming. In: Daniel, F., Wang, J., Weber, B. (eds.) Business Process Management - 11th International Conference, BPM 2013, Beijing, China, August 26-30, 2013. Proceedings. Lecture Notes in Computer Science, vol. 8094, pp. 113–129. Springer (2013). doi: 10.1007/978-3-642-40176-3_10, https://doi.org/10.1007/978-3-642-40176-3_10
71. Liu, R., Vaculín, R., Shan, Z., Nigam, A., Wu, F.Y.: Business artifact-centric modeling for real-time performance monitoring. In: Rinderle-Ma et al. [114], pp. 265–280. doi: 10.1007/978-3-642-23059-2_21, https://doi.org/10.1007/978-3-642-23059-2_21
72. López, M.T.G., Gasca, R.M., Rinderle-Ma, S.: Explaining the incorrect temporal events during business process monitoring by means of compliance rules and model-based diagnosis. In: Bagheri, E., Gasevic, D., Hallé, S., Hatala, M., Nezhad, H.R.M., Reichert, M. (eds.) 17th IEEE International Enterprise Distributed Object Computing Conference Workshops, EDOC Workshops, Vancouver, BC, Canada, September 9-13, 2013. pp. 163–172. IEEE Computer Society (2013). doi: 10.1109/EDOCW.2013.25, https://doi.org/10.1109/EDOCW.2013.25
73. Ly, L.T., Maggi, F.M., Montali, M., Rinderle-Ma, S., van der Aalst, W.M.P.: Compliance monitoring in business processes: Functionalities, application, and tool-support. Information Systems **54**, 209–234 (2015). doi: 10.1016/j.is.2015.02.007, https://doi.org/10.1016/j.is.2015.02.007
74. Ly, L.T., Rinderle-Ma, S., Knuplesch, D., Dadam, P.: Monitoring business process compliance using compliance rule graphs. In: Meersman, R., Dillon, T.S., Herrero, P., Kumar, A., Reichert, M., Qing, L., Ooi, B.C., Damiani, E., Schmidt, D.C., White, J., Hauswirth, M., Hitzler, P., Mohania, M.K. (eds.) On the Move to Meaningful Internet Systems: OTM 2011 - Confederated International Conferences: CoopIS, DOA-SVI, and ODBASE 2011, Hersonissos, Crete, Greece, October 17-21, 2011, Proceedings, Part I. Lecture Notes in Computer Science, vol. 7044, pp. 82–99. Springer (2011). doi: 10.1007/978-3-642-25109-2_7, https://doi.org/10.1007/978-3-642-25109-2_7
75. Maamar, Z., Faci, N., Sellami, M., Boukadi, K., Yahya, F., Barnawi, A., Sakr, S.: On business process monitoring using cross-flow coordination. Service Oriented Computing and Applications **11**(2), 203–215 (2017). doi: 10.1007/s11761-017-0206-0, https://doi.org/10.1007/s11761-017-0206-0
76. Maggi, F.M., Bose, R.P.J.C., van der Aalst, W.M.P.: Efficient discovery of understandable declarative process models from event logs. In: Ralyté et al. [110], pp. 270–285. doi: 10.1007/978-3-642-31095-9_18, https://doi.org/10.1007/978-3-642-31095-9_18
77. Maggi, F.M., Francescomarino, C.D., Dumas, M., Ghidini, C.: Predictive monitoring of business processes. In: Jarke, M., Mylopoulos, J., Quix, C., Rolland, C., Manolopoulos, Y., Mouratidis, H., Horkoff, J. (eds.) Advanced Information Systems Engineering - 26th International Conference, CAiSE 2014, Thessaloniki, Greece, June 16-20, 2014. Proceedings. Lecture Notes in Computer Science, vol. 8484, pp. 457–472. Springer (2014). doi: 10.1007/978-3-319-07881-6_31, https://doi.org/10.1007/978-3-319-07881-6_31
78. Maggi, F.M., Montali, M., Westergaard, M., van der Aalst, W.M.P.: Monitoring business constraints with linear temporal logic: An approach based on colored automata. In: Rinderle-Ma et al. [114], pp. 132–147. doi: 10.1007/978-3-642-23059-2_13, https://doi.org/10.1007/978-3-642-23059-2_13
79. Maggi, F.M., Westergaard, M., Montali, M., van der Aalst, W.M.P.: Runtime verification of ltl-based declarative process models. In: Khurshid and Sen [60], pp. 131–146. doi: 10.1007/978-3-642-29860-8_11, https://doi.org/10.1007/978-3-642-29860-8_11

80. Mandal, S., Hewelt, M., Weske, M.: A framework for integrating real-world events and business processes in an iot environment. In: Panetto, H., Debruyne, C., Gaaloul, W., Papazoglou, M.P., Paschke, A., Ardagna, C.A., Meersman, R. (eds.) On the Move to Meaningful Internet Systems. OTM 2017 Conferences - Confederated International Conferences: CoopIS, C&TC, and ODBASE 2017, Rhodes, Greece, October 23-27, 2017, Proceedings, Part I. Lecture Notes in Computer Science, vol. 10573, pp. 194–212. Springer (2017). doi: 10.1007/978-3-319-69462-7_13, https://doi.org/10.1007/978-3-319-69462-7_13

81. Marin, M., Hull, R., Vaculín, R.: Data centric BPM and the emerging case management standard: A short survey. In: La Rosa and Soffer [68], pp. 24–30. doi: 10.1007/978-3-642-36285-9_4, https://doi.org/10.1007/978-3-642-36285-9_4

82. Meroni, G.: Integrating the internet of things with business process management: A process-aware framework for smart objects. In: Loucopoulos, P., Nurcan, S., Weigand, H. (eds.) Proceedings of the CAiSE'2015 Doctoral Consortium at the 27th International Conference on Advanced Information Systems Engineering (CAiSE 2015), Stockholm, Sweden, June 11-12, 2015. CEUR Workshop Proceedings, vol. 1415, pp. 56–64. CEUR-WS.org (2015), http://ceur-ws.org/Vol-1415/CAISE2015DC07.pdf

83. Meroni, G., Baresi, L., Montali, M., Plebani, P.: Multi-party business process compliance monitoring through iot-enabled artifacts. Information Systems 73, 61 – 78 (2018). doi: 10.1016/j.is.2017.12.009, https://doi.org/10.1016/j.is.2017.12.009

84. Meroni, G., Baresi, L., Plebani, P.: Translating BPMN to E-GSM: specifications and rules. Tech. rep., Politecnico di Milano (2016), http://hdl.handle.net/11311/976678

85. Meroni, G., Di Ciccio, C., Mendling, J.: An artifact-driven approach to monitor business processes through real-world objects. In: Maximilien, E.M., Vallecillo, A., Wang, J., Oriol, M. (eds.) Service-Oriented Computing - 15th International Conference, ICSOC 2017, Malaga, Spain, November 13-16, 2017, Proceedings. Lecture Notes in Computer Science, vol. 10601, pp. 297–313. Springer (2017). doi: 10.1007/978-3-319-69035-3_21, https://doi.org/10.1007/978-3-319-69035-3_21

86. Meroni, G., Di Ciccio, C., Mendling, J.: Artifact-driven process monitoring: Dynamically binding real-world objects to running processes. In: Franch, X., Ralyté, J., Matulevicius, R., Salinesi, C., Wieringa, R. (eds.) Proceedings of the Forum and Doctoral Consortium Papers Presented at the 29th International Conference on Advanced Information Systems Engineering, CAiSE 2017, Essen, Germany, June 12-16, 2017. CEUR Workshop Proceedings, vol. 1848, pp. 105–112. CEUR-WS.org (2017), http://ceur-ws.org/Vol-1848/CAiSE2017_Forum_Paper14.pdf

87. Meroni, G., Montali, M., Baresi, L., Plebani, P.: Translating BPMN to E-GSM: proof of correctness. Tech. rep., Politecnico di Milano (2016), http://hdl.handle.net/11311/990248

88. Meroni, G., Plebani, P.: Artifact-driven monitoring for human-centric business processes with smart devices: Assessment and improvement. In: Carmona, J., Engels, G., Kumar, A. (eds.) Business Process Management Forum - BPM Forum 2017, Barcelona, Spain, September 10-15, 2017, Proceedings. Lecture Notes in Business Information Processing, vol. 297, pp. 160–176. Springer (2017). doi: 10.1007/978-3-319-65015-9_10, https://doi.org/10.1007/978-3-319-65015-9_10

89. Metzger, A., Leitner, P., Ivanovic, D., Schmieders, E., Franklin, R., Carro, M., Dustdar, S., Pohl, K.: Comparing and combining predictive business process monitoring techniques. IEEE Trans. Systems, Man, and Cybernetics: Systems 45(2), 276–290 (2015). doi: 10.1109/TSMC.2014.2347265, https://doi.org/10.1109/TSMC.2014.2347265

90. Meyer, A., Weske, M.: Activity-centric and artifact-centric process model roundtrip. In: Lohmann, N., Song, M., Wohed, P. (eds.) Business Process Management Workshops - BPM 2013 International Workshops, Beijing, China, August 26, 2013, Revised Papers. Lecture Notes in Business Information Processing, vol. 171, pp. 167–181. Springer (2013). doi: 10.1007/978-3-319-06257-0_14, https://doi.org/10.1007/978-3-319-06257-0_14

91. Miorandi, D., Sicari, S., De Pellegrini, F., Chlamtac, I.: Internet of things: Vision, applications and research challenges. Ad Hoc Networks 10(7), 1497–1516 (2012)

92. Montali, M., Maggi, F.M., Chesani, F., Mello, P., van der Aalst, W.M.P.: Monitoring business constraints with the event calculus. ACM TIST 5(1), 17:1–17:30 (2013). doi: 10.1145/2542182.2542199, http://doi.acm.org/10.1145/2542182.2542199

93. Motahari-Nezhad, H.R., Recker, J., Weidlich, M. (eds.): Business Process Management - 13th International Conference, BPM 2015, Innsbruck, Austria, August 31 - September 3, 2015, Proceedings, Lecture Notes in Computer Science, vol. 9253. Springer (2015). doi: 10.1007/978-3-319-23063-4, https://doi.org/10.1007/978-3-319-23063-4

94. Munoz-Gama, J.: Conformance Checking and Diagnosis in Process Mining - Comparing Observed and Modeled Processes, Lecture Notes in Business Information Processing, vol. 270. Springer (2016). doi: 10.1007/978-3-319-49451-7, https://doi.org/10.1007/978-3-319-49451-7

95. de Murillas, E.G.L., van der Aalst, W.M.P., Reijers, H.A.: Process mining on databases: Unearthing historical data from redo logs. In: Motahari-Nezhad et al. [93], pp. 367–385. doi: 10.1007/978-3-319-23063-4_25, https://doi.org/10.1007/978-3-319-23063-4_25

96. Musen, M.A.: The protégé project: a look back and a look forward. AI Matters 1(4), 4–12 (2015). doi: 10.1145/2757001.2757003, http://doi.acm.org/10.1145/2757001.2757003

97. Nambi, S.N.A.U., Sarkar, C., Prasad, R.V., Biswas, A.R.: A unified semantic knowledge base for iot. In: IEEE World Forum on Internet of Things, WF-IoT 2014, Seoul, South Korea, March 6-8, 2014. pp. 575–580. IEEE Computer Society (2014). doi: 10.1109/WF-IoT.2014. 6803232, https://doi.org/10.1109/WF-IoT.2014.6803232

98. Namiri, K., Stojanovic, N.: Pattern-based design and validation of business process compliance. In: Meersman, R., Tari, Z. (eds.) On the Move to Meaningful Internet Systems 2007: CoopIS, DOA, ODBASE, GADA, and IS, OTM Confederated International Conferences CoopIS, DOA, ODBASE, GADA, and IS 2007, Vilamoura, Portugal, November 25-30, 2007, Proceedings, Part I. Lecture Notes in Computer Science, vol. 4803, pp. 59–76. Springer (2007). doi: 10.1007/978-3-540-76848-7_6, https://doi.org/10.1007/978-3-540-76848-7_6

99. Narendra, N.C., Varshney, V.K., Nagar, S., Vasa, M., Bhamidipaty, A.: Optimal control point selection for continuous business process compliance monitoring. In: 2008 IEEE International Conference on Service Operations and Logistics, and Informatics. vol. 2, pp. 2536–2541 (Oct 2008). doi: 10.1109/SOLI.2008.4682963

100. Nezhad, H.R.M., Swenson, K.D.: Adaptive case management: Overview and research challenges. In: IEEE 15th Conference on Business Informatics, CBI 2013, Vienna, Austria, July 15-18, 2013. pp. 264–269. IEEE Computer Society (2013). doi: 10.1109/CBI.2013.44, https://doi.org/10.1109/CBI.2013.44

101. Nigam, A., Caswell, N.S.: Business artifacts: An approach to operational specification. IBM Systems Journal 42(3), 428–445 (2003). doi: 10.1147/sj.423.0428, https://doi.org/10.1147/sj.423.0428

102. Nooijen, E.H.J., van Dongen, B.F., Fahland, D.: Automatic discovery of data-centric and artifact-centric processes. In: La Rosa and Soffer [68], pp. 316–327. doi: 10.1007/978-3-642-36285-9_36, https://doi.org/10.1007/978-3-642-36285-9_36

103. Perera, C., Zaslavsky, A., Christen, P., Compton, M., Georgakopoulos, D.: Context-aware sensor search, selection and ranking model for internet of things middleware. In: 2013 IEEE 14th International Conference on Mobile Data Management. vol. 1, pp. 314–322 (Jun 2013). doi: 10.1109/MDM.2013.46

104. Perera, C., Zaslavsky, A.B., Christen, P., Georgakopoulos, D.: Context aware computing for the internet of things: A survey. IEEE Communications Surveys and Tutorials 16(1), 414–454 (2014). doi: 10.1109/SURV.2013.042313.00197, https://doi.org/10.1109/SURV.2013.042313.00197

105. Pérez, J., Arenas, M., Gutiérrez, C.: Semantics and complexity of SPARQL. In: Cruz, I.F., Decker, S., Allemang, D., Preist, C., Schwabe, D., Mika, P., Uschold, M., Aroyo, L. (eds.) The Semantic Web - ISWC 2006, 5th International Semantic Web Conference, ISWC 2006, Athens, GA, USA, November 5-9, 2006, Proceedings. Lecture Notes in Computer Science, vol. 4273, pp. 30–43. Springer (2006). doi: 10.1007/11926078_3, https://doi.org/10.1007/11926078_3

106. Pérez-Castillo, R., Weber, B., Pinggera, J., Zugal, S., de Guzmán, I.G.R., Piattini, M.: Generating event logs from non-process-aware systems enabling business process mining. Enterprise IS 5(3), 301–335 (2011). doi: 10.1080/17517575.2011.587545, https://doi.org/10.1080/17517575.2011.587545

107. Pesic, M., van der Aalst, W.M.P.: A declarative approach for flexible business processes management. In: Eder, J., Dustdar, S. (eds.) Business Process Management Workshops, BPM 2006 International Workshops, BPD, BPI, ENEI, GPWW, DPM, semantics4ws, Vienna, Austria, September 4-7, 2006, Proceedings. Lecture Notes in Computer Science, vol. 4103, pp. 169–180. Springer (2006). doi: 10.1007/11837862_18, https://doi.org/10.1007/11837862_18

108. Pesic, M., Schonenberg, H., van der Aalst, W.M.P.: DECLARE: full support for loosely-structured processes. In: 11th IEEE International Enterprise Distributed Object Computing Conference (EDOC 2007), 15-19 October 2007, Annapolis, Maryland, USA. pp. 287–300. IEEE Computer Society (2007). doi: 10.1109/EDOC.2007.14, https://doi.org/10.1109/EDOC.2007.14

109. Puccinelli, D., Haenggi, M.: Wireless sensor networks: applications and challenges of ubiquitous sensing. IEEE Circuits and Systems Magazine 5(3), 19–31 (2005). doi: 10.1109/MCAS.2005.1507522

110. Ralyté, J., Franch, X., Brinkkemper, S., Wrycza, S. (eds.): Advanced Information Systems Engineering - 24th International Conference, CAiSE 2012, Gdansk, Poland, June 25-29, 2012. Proceedings, Lecture Notes in Computer Science, vol. 7328. Springer (2012). doi: 10.1007/978-3-642-31095-9, https://doi.org/10.1007/978-3-642-31095-9

111. Reichert, M., Reijers, H.A. (eds.): Business Process Management Workshops - BPM 2015, 13th International Workshops, Innsbruck, Austria, August 31 - September 3, 2015, Revised Papers, Lecture Notes in Business Information Processing, vol. 256. Springer (2016). doi: 10.1007/978-3-319-42887-1, https://doi.org/10.1007/978-3-319-42887-1

112. Reichert, M., Weber, B.: Enabling Flexibility in Process-Aware Information Systems - Challenges, Methods, Technologies. Springer (2012). doi: 10.1007/978-3-642-30409-5, https://doi.org/10.1007/978-3-642-30409-5

113. Richardson, L., Ruby, S.: RESTful web services - web services for the real world. O'Reilly (2007)

114. Rinderle-Ma, S., Toumani, F., Wolf, K. (eds.): Business Process Management - 9th International Conference, BPM 2011, Clermont-Ferrand, France, August 30 - September 2, 2011. Proceedings, Lecture Notes in Computer Science, vol. 6896. Springer (2011). doi: 10.1007/978-3-642-23059-2, https://doi.org/10.1007/978-3-642-23059-2

115. Roussos, G., Kostakos, V.: Rfid in pervasive computing: state-of-the-art and outlook. Pervasive and Mobile Computing 5(1), 110–131 (2009)

116. Rozinat, A., van der Aalst, W.M.P.: Conformance checking of processes based on monitoring real behavior. Information Systems 33(1), 64–95 (2008). doi: 10.1016/j.is.2007.07.001, https://doi.org/10.1016/j.is.2007.07.001

117. Russell, N., Hofstede, A.H.M.T., Mulyar, N.: Workflow controlflow patterns: A revised view. Tech. Rep. BPM-06-22, BPM Center Report, BPMcenter.org (2006)

118. Santos, E.A.P., Francisco, R., Vieira, A.D., Loures, E.D.F.R., de Paula, M.A.B.: Modeling business rules for supervisory control of process-aware information systems. In: Daniel, F., Barkaoui, K., Dustdar, S. (eds.) Business Process Management Workshops - BPM 2011 International Workshops, Clermont-Ferrand, France, August 29, 2011, Revised Selected Papers, Part II. Lecture Notes in Business Information Processing, vol. 100, pp. 447–458. Springer (2011). doi: 10.1007/978-3-642-28115-0_42, https://doi.org/10.1007/978-3-642-28115-0_42

119. Schmidt, W., Fleischmann, A.: Business process monitoring with S-BPM. In: Fischer, H., Schneeberger, J. (eds.) S-BPM ONE - Running Processes, 5th International Conference, S-BPM ONE 2013, Deggendorf, Germany, March 11-12, 2013. Proceedings. Communications in Computer and Information Science, vol. 360, pp. 274–291. Springer (2013). doi: 10.1007/978-3-642-36754-0_18, https://doi.org/10.1007/978-3-642-36754-0_18

120. Sebahi, S.: Monitoring business process compliance : a view based approach. (Monitoring de la conformité des processus métiers : approche à base de vues). Ph.D. thesis, Claude Bernard University Lyon 1, France (2012), https://tel.archives-ouvertes.fr/tel-00866483

121. Senderovich, A., Rogge-Solti, A., Gal, A., Mendling, J., Mandelbaum, A.: The ROAD from sensor data to process instances via interaction mining. In: Nurcan, S., Soffer, P., Bajec, M., Eder, J. (eds.) Advanced Information Systems Engineering - 28th International Conference,

CAiSE 2016, Ljubljana, Slovenia, June 13-17, 2016. Proceedings. Lecture Notes in Computer Science, vol. 9694, pp. 257–273. Springer (2016). doi: 10.1007/978-3-319-39696-5_16, https://doi.org/10.1007/978-3-319-39696-5_16

122. Stertz, F., Mangler, J., Rinderle-Ma, S.: Nfc-based task enactment for automatic documentation of treatment processes. In: Reinhartz-Berger, I., Gulden, J., Nurcan, S., Guédria, W., Bera, P. (eds.) Enterprise, Business-Process and Information Systems Modeling - 18th International Conference, BPMDS 2017, 22nd International Conference, EMMSAD 2017, Held at CAiSE 2017, Essen, Germany, June 12-13, 2017, Proceedings. Lecture Notes in Business Information Processing, vol. 287, pp. 34–48. Springer (2017). doi: 10.1007/978-3-319-59466-8_3, https://doi.org/10.1007/978-3-319-59466-8_3

123. Swenson, K.: Taming the unpredictable: real world adaptive case management: case studies and practical guidance. Future Strategies Inc. (2011)

124. Teixeira, T., Hachem, S., Issarny, V., Georgantas, N.: Service oriented middleware for the internet of things: A perspective. In: Towards a Service-Based Internet. Springer (Jan 2011). doi: 10.1007/978-3-642-24755-2_21, http://dx.doi.org/10.1007/978-3-642-24755-2_21

125. Thullner, R., Rozsnyai, S., Schiefer, J., Obweger, H., Suntinger, M.: Proactive business process compliance monitoring with event-based systems. In: Workshops Proceedings of the 15th IEEE International Enterprise Distributed Object Computing Conference, EDOCW 2011, Helsinki, Finland, August 29 - September 2, 2011. pp. 429–437. IEEE Computer Society (2011). doi: 10.1109/EDOCW.2011.22, https://doi.org/10.1109/EDOCW.2011.22

126. Uschold, M., Gruninger, M.: Ontologies: principles, methods and applications. Knowledge Eng. Review 11(2), 93–136 (1996). doi: 10.1017/S0269888900007797, https://doi.org/10.1017/S0269888900007797

127. Vandermerwe, S., Rada, J.: Servitization of business: Adding value by adding services. European Management Journal 6(4), 314 – 324 (1988). doi: http://dx.doi.org/10.1016/0263-2373(88)90033-3, http://www.sciencedirect.com/science/article/pii/0263237388900333

128. Vanhatalo, J., Völzer, H., Koehler, J.: The refined process structure tree. Data Knowl. Eng. 68(9), 793–818 (2009). doi: 10.1016/j.datak.2009.02.015, https://doi.org/10.1016/j.datak.2009.02.015

129. Wang, W., De, S., Tönjes, R., Reetz, E.S., Moessner, K.: A comprehensive ontology for knowledge representation in the internet of things. In: Min, G., Wu, Y., Liu, L.C., Jin, X., Jarvis, S.A., Al-Dubai, A.Y. (eds.) 11th IEEE International Conference on Trust, Security and Privacy in Computing and Communications, TrustCom 2012, Liverpool, United Kingdom, June 25-27, 2012. pp. 1793–1798. IEEE Computer Society (2012). doi: 10.1109/TrustCom.2012.20, https://doi.org/10.1109/TrustCom.2012.20

130. Weber, B., Pinggera, J., Neurauter, M., Zugal, S., Martini, M., Furtner, M., Sachse, P., Schnitzer, D.: Fixation patterns during process model creation: Initial steps toward neuroadaptive process modeling environments. In: Bui, T.X., Jr., R.H.S. (eds.) 49th Hawaii International Conference on System Sciences, HICSS 2016, Koloa, HI, USA, January 5-8, 2016. pp. 600–609. IEEE Computer Society (2016). doi: 10.1109/HICSS.2016.81, https://doi.org/10.1109/HICSS.2016.81

131. Weerdt, J.D., vanden Broucke, S.K.L.M., Vanthienen, J., Baesens, B.: Active trace clustering for improved process discovery. IEEE Trans. Knowl. Data Eng. 25(12), 2708–2720 (2013). doi: 10.1109/TKDE.2013.64, https://doi.org/10.1109/TKDE.2013.64

132. Weske, M.: Business Process Management - Concepts, Languages, Architectures, 2nd Edition. Springer (2012). doi: 10.1007/978-3-642-28616-2, https://doi.org/10.1007/978-3-642-28616-2

133. Wombacher, A.: How physical objects and business workflows can be correlated. In: Jacobsen, H., Wang, Y., Hung, P. (eds.) IEEE International Conference on Services Computing, SCC 2011, Washington, DC, USA, 4-9 July, 2011. pp. 226–233. IEEE Computer Society (2011). doi: 10.1109/SCC.2011.24, https://doi.org/10.1109/SCC.2011.24

134. Xu, Y., Zhang, C., Ji, Y.: An upper-ontology-based approach for automatic construction of IOT ontology. IJDSN 10 (2014). doi: 10.1155/2014/594782, https://doi.org/10.1155/2014/594782

135. Zhu, D., Cheng, B., Zhang, Y., Chen, J.: Future service provision: Towards a flexible hybrid service supporting platform. In: 5th IEEE Asia-Pacific Services Computing Conference, APSCC 2010, 6-10 December 2010, Hangzhou, China, Proceedings. pp. 226–233. IEEE Computer Society (2010). doi: 10.1109/APSCC.2010.51, https://doi.org/10.1109/APSCC.2010.51

Printed in the United States
By Bookmasters